Jiří Pokorný
and Tsu-Ming Wu

BIOPHYSICAL ASPECTS OF COHERENCE AND BIOLOGICAL ORDER

Springer-Verlag Berlin Heidelberg GmbH

Biophysical Aspects of Coherence and Biological Order

JIŘÍ POKORNÝ
AND TSU-MING WU

With 60 Figures
and 6 Tables

Springer

Co-edition with Academia, Publishing House of the Academy of Sciences of the Czech Republic
Academia has exclusive distribution rights for the Czech Republic, the Slovak Republic, Russia and the other Eastern European countries

ISBN 978-3-662-03549-8

Die Deutsche Bibliothek - CIP-Einheitsaufnahme
Pokorný, Jiří:
Biophysical aspects of coherence and biological order: 6 tables/Jiří Pokorný; Tsu-Ming Wu.

ISBN 978-3-662-03549-8 ISBN 978-3-662-03547-4 (eBook)
DOI 10.1007/978-3-662-03547-4

DBN: 95.392137.9
P SG: 32;33
110

Cover design: Oleg Man, Czech Republic
Typesetting: Camera-ready by Jiří Pokorný

SPIN 10678229 31/3137 5 4 3 2 1 0 - Printed on acid free paper

Contents

Preface

Biological order is a phenomenon which has not been satisfactorily explained yet. Various components join together to form larger, stable structures according to certain rules holding for self-assembly. Their new properties cannot be predicted from the characteristics of their parts. A large amount of biochemical reactions and biophysical processes including an inconceivable number of various entities (molecules, structures, organelles, etc.) are coordinated simultaneously and successively giving rise to a well organized dynamic system whose dimensions are much greater than the "action radii" of chemical reactions, of electrostatic and quasi-electrostatic interactions screened by moving ions, or of van der Waals–London dispersion interactions. From the point of view of thermodynamics biological systems are open systems far from thermodynamic equilibrium. Order is supposed to depend on certain long range biophysical mechanisms using energy.

Before discussing the mechanisms which we presume to underlie biological order we have to emphasize some typical features of the biological mass: extraordinary polarity and polarizability, non-linearity, and energy supply. Without going into details we will show the consequences of these properties. Even thermal vibrations in biological molecules and structures which carry electric charges or are strongly polarized are accompanied by polarization waves generating an electromagnetic field capable of mediation of long range interactions.

In a non-linear system a small disturbance can result in great changes of the system at a later time. Oscillations are connected with spectral transformation of energy. Energy supplied into the non-linear systems (in biological systems energy from metabolic sources, e.g. from the hydrolysis of adenosine triphosphate) need not be thermalized immediately and thus nonequilibrium thermodynamic distribution of energy may be created. The system can be shifted far from thermodynamic equilibrium. In vibration systems energy may be accumulated predominantly in certain vibration modes. Spectral energy transformation in oscillation systems is described by the Manley–Rowe equations (originally derived for classical oscillators), which represent the expectation values for quantum oscillators too.

The book contains the theory and model calculations of a biophysical mechanism based on energy condensation in the vibration systems. As a model non-linear system we use the Fröhlich vibration system to analyse the basic properties of coherent states. Computer models make it possible to reveal vibration modes with nonequilibrium energy content, to assess the effects of the non-linear coupling to the surroundings and of the extinction of energy condensation, and to describe interaction and information transfer (for instance, between excited molecules). The electromagnetic field generated by the assumed coherent vibrations in molecules or structures in a cell can act up to a distance of the order of 0.1–1 μm (with interference patterns of similar dimensions). Therefore, the electromagnetic field is a long range field in the mesoscopic cellular world but a short range field for macroscopic measurement systems.

The cytoskeleton is a structure that satisfies the requirements for excitation of the coherent vibrations. The important polymers of the cytoskeleton – the microtubules and the actin filaments – are polar and obtain a large amount of energy from the hydrolysis of energy rich nucleotides. The cytoskeleton exerts forces and generates movements in the cell as its main organizer.

The vibration systems in biological molecules and structures can be strongly influenced by the mechanism of phosphorylation. Protein phosphorylation is a general controlling mechanism in the cells. Phosphorylated residues of amino acids have large masses with separated charges and represent regions of the strong Coulomb coupling of proteins to their surroundings. We present the hypothesis that phosphorylation alters the frequencies of vibration modes and causes strong damping of vibrations in proteins. Disturbances of coherence may explain certain aspects of cancer cell behaviour.

The text of the book is accessible to a wide community of readers. The necessary physical, mathematical and chemical items are explained in the appendices. Only the part dealing with the derivation of the Fröhlich rate equation based on the Green functions is not easy for the readers who are not familiar with this technique. This part was written by Professor T.-M. Wu. The transition to the description of the Hamiltonian by the Green functions is based on the general rules derived by Abrikosov et al. in the book "Quantum Field Theoretical Methods in Statistical Physics". Nevertheless, the part provides an alternative derivation of the Fröhlich rate equation.

J. Pokorný

1. Biological Order

1.1 Biophysical Mechanisms

Molecular and biochemical research is in an explosive phase and achieved a great success in biology. The structure and configuration of biological macromolecules (especially of deoxyribonucleic acid – DNA, ribonucleic acid – RNA, and proteins) have been disclosed and the cycles and chains of biochemical reactions have been described (e.g. the chain of reactions producing adenosine triphosphate – ATP with a dominant role of the citric acid cycle). Research in biology, molecular biology, and biochemistry may be characterized as a study of the composition, the space configuration and conformation of biological macromolecules and structures, and chemical reactions between them. Regardless of the great discoveries and their fundamental significance for biology this approach can describe only certain aspects of biological systems.

The enormous complexity and organization of biological systems, not only at the macroscopic level, but especially at the microscopic molecular level, is stunning. Countless amounts of various entities (molecules, molecular structures, organelles, etc.) interact, cooperate and are bound together in an ordered way simultaneously and successively in time in a large space whose dimensions are many orders greater than the "action radii" of chemical reactions: a well organized dynamic system is created. Order in biological systems is discussed in the following way [1]:

It is common experience that things left to themselves eventually become disordered: buildings crumble, dead organisms become oxidized, and so on. This general tendency is expressed in the second law of thermodynamics, which states that the degree of disorder in the universe (or in any isolated system in the universe) can only increase.

The puzzle is that living organisms maintain, at every level, a very high degree of order; and as they feed, develop, and grow, they appear to create this order out of raw materials that lack it. Order is strikingly apparent in large structures such as a butterfly wing or an octopus eye, in subcellular structures such as a mitochondrion or cilium, and in the shape and arrangement of molecules from which these structures are built.

Excluding the question of order and disorder in the universe we have to emphasize that any order may be established by the organized activity and "work" which requires energy. Therefore, order in biological systems has to be connected with energy supply [1]:

Living cells are highly ordered and must create order within themselves in order to survive and grow. This is thermodynamically possible only because of continual input of energy, part of it is released from the cell to its environment as heat.

Photosynthetic organisms use solar electromagnetic energy and convert it into the energy of chemical bonds. All the "nonphotosynthetic" organisms provide energy by eating plants or by eating other animals (whose feeding chains begin on photosynthetic organisms). The organic molecules made by plant cells provide both building blocks as well as fuel for organization of order. Nevertheless, energy supply by itself cannot lead to and maintain any order. Energy has to be used in an order inducing mechanism. This idea is discussed e.g. in [1]:

It is important to note that the cell will achieve nothing by producing heat unless the heat-generating reactions are directly linked with the processes that generate molecular order in the cell.

There is no doubt that mechanisms of biological order are far from random movements and collisions. The experimental findings suggest that interactions of biophysical nature based on energy supply are responsible for these effects.

Before we start with analysis of the biophysical mechanisms of order we will mention some concepts of chemical reactions. The concept of affinity is widely used in chemistry to assess the strength and the specificity of the interaction between entities (e.g. between molecules). The equilibrium constant (which is sometimes termed the affinity constant) for the reaction employed as a measure of the affinity is given by the ratio of the association rate constant to the dissociation rate constant (Appendix 13.14). The equilibrium state of a reaction is established when the association rate equals the dissociation rate of the reaction in question. The greater the strength of the binding the larger the value of the equilibrium constant. The equilibrium constant of a reaction in which two components bind to each other is related directly to the standard free energy change for the binding. Nevertheless, the affinity represents an integral effect of many different mechanisms and processes. Before binding the molecules have to recognize one another and to move to the same place into close contact, rotate and adjust a correct position to bring the interacting surfaces together. It is generally accepted that besides the thermal motion some force mechanisms can bring the interacting entities together, e.g. the mechanisms based on ionic and van der Waals forces, on hydrogen bonds, and on forces acting on the non-polar entities caused by their hydrophobic expulsion from water. When the entities are in a close contact they can bind to each other. The strength of the binding depends on the nature of the bond. We have to emphasize that the affinity as an empirical integral quantity – under non-standard conditions – does not make possible to determine the mechanisms of interaction, their nature, the non-thermal transport of reaction components, and to distinguish coherent processes from incoherent processes. Forces of various origin can have strong effects on biochemical reactions.

We would mention some typical forces that can operate in living systems. These forces are surveyed e.g. in [1,2]. Van der Waals forces originate from the movement of charges in atoms. The movement of negative electrons around a positive nucleus gives rise to a fluctuating electromagnetic field. Similar mechanisms can exist in molecules. Electrostatic double layer forces are always present between charged surfaces in a liquid medium with mobile ionic charges. These forces depend on the overlap of the diffuse layers containing counterions compensating the charges on the surfaces of both layers. Hydration forces are present between hydrophilic surfaces such as surfactant headgroups. Steric-protrusion forces depend on a decreased mobility of molecules at the interface. The decreased mobility perpendicular to the surface gives rise to a steric type of repulsion. Hydrophobic attraction forces are the long range forces. Their mechanism includes repulsion between hydrophobic groups and water molecules. A variety of polymer induced forces including steric repulsion, bridging attraction, and other forces leading to hydrogen bonds and ionic bridges exist in organic and inorganic matter.

A matter of importance is the following: Can these forces organize biological systems? Can they transform energy supplied e.g. from the adenosine triphosphate or guanosine triphosphate molecules into useful work? Parsegian discusses this point in [3]:

Once you crack the nanometer barrier, you see forces, dynamics and mechanics – often excitingly unlike their atomic or colloidal or macroscopic analogs. Still, their idiosyncrasies can be countenanced and described in classical physical language. On this size scale there is even a clear and new link between physical work and reaction chemistry. How does the work of ATP hydrolysis couple to protein to create physical force? Burried inside all cartoons and chemistry of bioenergetics is a "then-comes-a-miracle" step from chemical energy to physical force.

We entertain an idea that certain long range interaction forces (not specified yet) act in biological systems causing cooperation among a great number of entities in a large area. Cooperative phenomena in living matter are conditioned by mass transport, energy and information transfer. Cooperation gives rise to coherent dynamic regions – large subsystems on different levels – and to creation of a hierarchical structure of biological systems. Coherence is generally accepted by the scientific community as a fundamental physical principle (e.g. underlying lasers and masers). The mechanisms of coherent interactions and cooperation enable selforganization of biological systems. Selforganization is assumed to be one of the essential principles in the evolution of biological systems and in the processes of life as is given in [4,5]. Therefore, we will focus our attention on the mechanisms using energy in living matter. Metabolic energy is supplied to biological systems and a part of it is released as heat – a waste product – to their environment. From the point of view of thermodynamics, biological systems are open systems and as they are ordered they should be far from thermodynamic equilibrium.

In connection with order we have to assess another important question concerning linearity of biological systems. High electric field in membranes, proteins, and subcellular structures can alter their original linear properties into non-linear properties. Many molecules and structures may be non-linear even in thermodynamic equilibrium. Therefore, non-linearities in biological systems have to be included in any analysis of biological order.

The idea of coherence in biological systems originally emanated from the distinguished Professor Herbert Fröhlich F.R.S. Professor Fröhlich considered extraordinary dielectric and polarization properties of living matter. Vibrations in polar systems are accompanied by polarization waves generating an electromagnetic field which can mediate a long range interaction and play a role of the ordering agent. (In thermodynamic equilibrium the interaction might be regarded as the London dispersion interaction.) We have to point out that random vibrations are an omnipresent kinetic property of matter. As a further step Fröhlich applied the well known principles of non-linear physics to biology. Non-linearity in an oscillating system enable energy transfer between different spectral lines. Frequency mixing and energy transfer among the spectral lines (among vibration modes) occur in non-linear classical oscillators as well as in anharmonic quantum oscillators. Energy may be channelled from the high frequency oscillators downwards to the low frequency oscillators and upwards too. The spectral energy transformation is a general principle of non-linear systems and a condition for energy condensation. Energy supplied to a non-linear system is not immediately thermalized but condenses in certain spectral lines (only some degrees of freedom are excited while the system as a whole seems to remain in thermodynamic equilibrium). Energy distribution in an excited non-linear system may differ from the thermodynamic equilibrium distribution fundamentally. High excitation of some spectral lines leads to coherence of vibrations, i.e. according to the classical description well defined relationships of the amplitude and phase of vibrations are maintained in space and time. These ideas are the fundamentals of the Fröhlich theory explaining the mechanism of biological order. The coherent endogenous electric field of oscillating nature is assumed to be the key factor of organization. The basic ideas of this general Fröhlich hypothesis are given e.g. in [6–8] and surveyed and discussed in [9].

On the basis of the general principles Fröhlich derived a rate equation using a special type of interaction of a vibration system with its surroundings (i.e. with a heat bath). This interaction enables energy transfer between two vibration modes with different frequencies and the heat bath balances the difference in energy quanta [7,8]. As a result the Fröhlich system behaves like a non-linear system – a system with non-linear interaction with the heat bath.

We have to emphasize that the Fröhlich rate equation is a mathematical expression of only one possible non-linear mechanism of biological order. A wide diversity of non-linear mechanisms may exist in biological molecules, structures, and cells leading to generation of coherent or quasi-coherent electromagnetic fields, excitation of solitary waves etc. The concept of coherence in biological systems cannot be confined only to harmonic oscillations keeping certain phase relations. Nevertheless, in this book we will focus our attention to the mechanisms of coherent oscillations and generation of an endogenous electromagnetic field.

The range of interaction mediated by the endogenous coherent electromagnetic field whose frequency and energy correspond to the similar quantities of excited oscillators or vibration modes is not only several orders greater than the "action radii" of chemical bonds but also than the range of the interaction of the static electric field, which is screened by mobile ions (Appendix 13.13). A protein may be selectively attracted to the locus of its reaction from a great distance; e.g. the activator or repressor of gene expression may easily "find" the interaction domain on the DNA chain. Vos et al. [10] assumed that the specific function of protein macromolecules relies on their internal nuclear dynamics which allow them to adopt conformations which are different in structure and properties.

Blumenfeld [11] discussed coherent nuclear vibrations in terms of their important role in the classical kinetics of chemical reactions:

A substrate-product transformation requires the overcoming of a certain activation barrier. The excitation of coherent nuclear vibration, i.e., the excitation of the specific degree of freedom coinciding with the reaction coordinate (here, as a matter of fact, lies the enzyme specificity) is equivalent to the creation of the inverse population at the system's higher and lower levels. In other words, the specific degree of freedom has a very high local temperature. The probability of its being "changed" for non-coherent vibrations of the nuclear subsystem is negligibly small, and the most probable relaxation path includes the chemical substrate-product transformation. Chemical transformation is thus realized at a very high local temperature, the effective kT value is much higher than the reaction activation barrier, which under these conditions cannot be regarded as an obstacle.

The Fröhlich hypothesis of non-thermal excitations and coherence opened a new area for progress in understanding of life. The enormous findings in molecular biology, biochemistry, and physical chemistry laid the foundations for the study of concrete biophysical mechanisms of coherent states. The structures that are subject to coherent vibrations can be found.

The coherent systems generating an endogenous electric field are also interesting from the point of view of the information theory. States with condensed energy contain information. The amount of information depends on the "distance" of the system from thermodynamic equilibrium, i.e. on the condensed energy and its distribution. The greater the condensed energy, the smaller the probability of the state and the greater its information content. Information transfer in biological systems can be mediated by the endogenous electromagnetic field too.

1.2 Discussion about Experiments

First of all we will discuss the prerequisites for coherent state excitation. We have to assess the possibility of synphase nuclear oscillations in biological molecules and structures and of non-linear interactions in the system. The experimental findings of Vos et al. [10] yield some data to this point. They irradiated proteins forming bacterial photosynthetic reaction centres by 40 and 80 fs pulses in the near infrared region (850–900 nm). After 0.1 ps waiting time they received the stimulated emission at some discrete wavelengths in the range from 900 to 960 nm (where no spectral overlap with the ground state absorption exists). The stimulated emission was modulated by coherent signals (of sinusoidal form) with the wavelength of 13 μm and of 66 μm and by coherent signals with the combination frequencies. We may conclude that coherent nuclear oscillations can be excited in the measured protein reaction centres which are non-linear.

A fundamental feature of coherent vibration states in polar molecules and structures is generation of the electromagnetic field with a dominant electric component. Measurements of the endogenous electromagnetic field in biological systems is of the utmost importance regardless of the mechanisms that underlie its generation. Some measurement methods are based on the force effects of the endogenous electromagnetic field. These methods should yield certain average values of the force exerted by a highly inhomogeneous field. For instance, experiments based on the cell–cell interactions and on cell–substrate interactions support the theory of the endogenous electromagnetic field generated by cells. Before surveying the main items of the published data on interactions between erythrocytes and between leukocytes we will quote some general molecular biology data on adhesion between cells from [1]:

To form an anchoring junction, cell must first adhere. A bulky cytoskeletal apparatus must then be assembled around the molecules that directly mediate the adhesion. The result is a well-defined structure – a desmosome, a hemidesmosome, or an adherens or septate junction – that is easily identified in the electron microscope. Indeed, electron microscopy provided the basis for the original classification of cell junctions. In the early stages of development of a cell junction, however, before the cytoskeletal apparatus has assembled, and especially in embryonic tissues, the cells often adhere to one another without clearly displaying these characteristic structures: in the electron microscope one may simply see two plasma membranes separated by a small gap of a definite width.

Therefore, it seems reasonable to accept an idea that the cell adhesion has two phases: in the first phase adhesion is not dependent on chemical bonds and in the second phase the chemical bonds are organized.

In [1] we may find further description of mechanisms supporting this idea, e.g. a concept of selective adhesion:

But the cells, as they accumulate, do not simply remain passively stuck together as a disorderly pile; instead, we shall see, the tissue architecture is actively maintained by selective adhesions that cell makes and progressively adjust. Thus if cells of different embryonic tissue are artificially mingled, they will often spontaneously sort out to restore a more normal arrangement...

A still more striking example of the same phenomenon is seen when dissociated cells from two embryonic vertebrate organs such as liver and retina are mixed together and artificially formed into pellet: the mixed aggregates gradually sort out according to their organ of origin. Similarly, disaggregated cells are found to adhere more readily to aggregates of their own organ than to aggregates of other organs. Evidently there are cell-cell recognition systems that make cells of the same differentiated tissue preferentially adhere to one another; these adhesive preferences are presumably important in stabilizing tissue architecture.

We will discuss cell adhesion based on molecular mechanisms. A general concept of adhesion is discussed e.g. in [12]. The ligand-receptor interaction depends on the electrostatic and steric repulsion and on van der Waals and hydrophobic attraction giving rise to a typical curve of the interaction energy with a minimum at a distance of several nanometers. The selective cell–cell adhesion is generally explained by two classes of the adhesion molecules, one Ca^{2+} dependent and the other Ca^{2+} independent [1]. Cadherins responsible for the Ca^{2+} dependent cell–cell adhesion have extracellular part folded in five domains each containing about 100 amino acids residues. The distance at which two cells are bound together is of about 10 nm. The Ca^{2+} independent cell–cell adhesion is mediated by the immunoglobulin superfamily of proteins which protrude from the membrane approximately to the same distance as cadherins. If the cells are bound to the extracellular matrix the distance between them may be greater. (The cells bind the extracellular matrix and respond to it by integrins.) Nevertheless, the cells recognize one another at a distance greater than the distance corresponding to the adherence based on chemical bonds of the adhesion molecules. The Brownian motion transports the cell in the suspension but with no selectivity. Explaining the long range interaction by van der Waals forces or London dispersion forces seems also to be rather difficult on account of the interaction distance of the order of 0.1–1 μm and of the selectivity in the first phase of adhesion. Even adjustment of the cells to right positions in the tissue is done before the chemical binding. These findings suggest that the first phase of adhesion is dependent on the long range cooperation mediated by some kind of a force field connected with cells.

Rowlands et al. [13–23] measured interaction forces among erythrocytes which appear to carry out the Brownian motion until they attach to each other forming rouleaus having appearance of a stack of coins. Investigating the motion of erythrocytes in terms of the Smoluchowski theory (which permits quantitative evaluation) Rowlands found out that interaction forces act between them up to a distance of about 1 μm. He observed that normal red blood cells were subject to attractive forces. By contrast, the cells fixed in glutardehyd, the cells at pH of 6.3, the metabolically depleted cells, and passive spheres do not attract one another. The metabolically depleted cells after repletion with adenosine and the cells, which were at pH of 6.3, after adjusting normal pH value of about 7.6 restore the attractive interaction. Nevertheless, the interaction is conditioned by the presence of certain extended macromolecules (e.g. of fibrinogen, polyvinylpyrrolidone, Dextran 70, etc.) in the solution in which the erythrocytes are suspended. The role of these macromolecules in erythrocyte interaction is not known. We have to mention some characteristic features of the erythrocyte outer surface. Two major proteins exposed on the outer surface of the human red blood cells are glycophorin and band 3 protein. Glycophorin is a small single-pass transmembrane glycoprotein (which has 131 amino acid residues) whose external part carries all of the carbohydrates and most of the negative charge of the surface. The function of glycophorin remains unknown. Band 3 protein is a multipass membrane protein which allows the bicarbonate (HCO_3^-) to cross the membrane and increases the amount of CO_2 which is removed from the blood [1]. It seems rather difficult to connect functions of these proteins with rouleau formation, for instance asuming that they are involved in the binding through an extracellular linker molecule. Erythrocytes have a shape of a biconcave circular disc. Even if their first point of touch is at the circumference they proceed to adjust the correct "stack of coins" position. The first phase of interaction starts at a distance several orders greater than London–van der Waals interactions.

We examined CD4 lymphocytes (CD4 Ly) a subpopulation of T lymphocytes. Lymphocytes belong to a class of the white blood cells. They are responsible for immune specificity. Two types of immune responses are mediated by T lymphocytes and B lymphocytes. B lymphocytes, which in mammals develop in the adult bone marrow or in the fetal liver, produce antibodies. T lymphocytes, which develop in the thymus, are involved in cell mediated immune responses. T cells contain cytotoxic T cells, which kill infected cells, and helper T cells, which enhance responses of other white blood cells. A co-receptor CD4 protein is expressed on helper T cells and mediates the strong binding to a family of proteins called class II major histocompatibility complexes (CD4 is also a receptor for HIV).

CD4 Ly were prepared from venous blood of healthy humans and cancer patients [24,25]. CD4 Ly were suspended in saline solution. In the presence of antigen CD4 Ly cells taken from immunized individuals (e.g. from cancer patients) exhibit decreased adherence to solid state surfaces. The majority of cells do not adhere. By contrast, these cells attracted one another and formed clusters in more than 60 % of all the cases. CD4 Ly from immunized individuals with no antigen in suspension adhered and did not form clusters as well as CD4 Ly taken from healthy humans (in this case with no antigen and with antigen in suspension).

We compare our experimetal findings with the data about lymphocytes given in [1] and [26]. The lymphocyte function is associated with antigen receptors and β_2 integrins. Integrins are composed of two noncovalently associated transmembrane glycoproteins and their globular head with the binding site is at a distance of about 20–30 nm from the lipid bilayer. The normal function of integrins is the binding to the extracellular matrix. By contrast, the β_2 integrins mainly mediate cell–cell, rather than cell–extracellular matrix interactions, by binding to specific ligands on another cell, such as an endothelial cell forming lining in a blood vessel; the lymphocyte attaches firmly and crosses the lining. Prototype characteristics of integrins is the divalent cation dependence of their adhesive interactions (i.e. dependence on Ca^{2+}, Mn^{2+}, Mg^{2+}). For instance, direct relationship between the extracellular Ca^{2+} binding and stimulation of $\alpha_L\beta_2$ integrin adhesiveness of T lymphocytes may be found. The integrins on blood cells often have to be activated before they can mediate cell adhesion. The mechanisms activating the integrins are largely unknown. Under the resting conditions the leukocyte integrins are randomly and homogeneously distributed throughout the cell surface. After activation the integrins make aggregates and clusters which is considered as an important step for efficient β_2 integrin mediated adhesive interaction. An important property of integrins is also their signal transduction capacity (from the outside into the cell and from the inside out of the cell).

The T cell receptors are composed of two polypeptide chains tightly associated in the plasma membrane with a number of proteins involved in passing the signal from the receptor to the cell interior. The initial activation of a

T cell usually occurs when it recognizes antigen bound to a major histocompatibility complex molecule on the surface of an antigen presenting cell. This mechanism is fundamental to the function of helper T cell. The T lymphocyte helper function is absolutely dependent on adhesive mechanism.

The integrins mediate adhesion at a distance of the order of about 10 nm similarly to the cadherins and the immunoglobulins. We do not assume that in our experiments the extracellular matrix capable of mediating long range interaction of CD4 Ly can be formed (if it can be formed at all as the cells are suspended in the saline solution) immediately after cells and antigen are given into the saline solution. Therefore, the long range interaction between CD4 Ly forming or organizing clusters can be hardly assigned to the extracellular matrix mechanism.

Adhesion to a solid state surface may be discussed in terms of the extracellular matrix too. Cells grown on a culture dish (e.g. fibroblasts) do not adhere to the surface directly. Most of their cell surface is separated from the substratum by a gap of more than 50 nm; but at focal contacts (adhesion plaques) this gap is reduced to 10–15 nm where the plasma membrane is attached by transmembrane linker glycoproteins to the extracellular matrix that has become adsorbed to the substratum [1]. In our case the CD4 Ly cells forming clusters do not adhere to the substratum and, therefore, it seems difficult to accept the extracellular matrix mechanism for explanation of the adhesion alterations.

We observed altered adhesion to solid state surfaces and the cluster formation in CD4 Ly; either effect can be explained in terms of the interaction mediated by a cell generated electromagnetic field. We assume that antigen effect results in important changes in coherent vibrations, especially in the strong damping of the vibrations. In a transient period after giving antigen into the suspension the vibrations may be damped only in certain regions and the remaining parts of the cell can oscillate with no damping. These oscillation "spots" can have random distribution. If the distribution is far from uniform a balance given by the condition of minimum energy is disturbed. A strong electromagnetic field may be generated (much greater than the field before antigen was bound) which mediates strong long range interactions between the cells and may be a factor organizing cell-cell adhesion. In the final phase adhesion might be mediated by integrins. If a steady state is reached the whole system of vibration is damped, the cell generated electromagnetic field is weak, interaction with the surface charge of the solid state surface is diminished, and the cell does not adhere.

Pohl et al. observed dielectrophoretic effects of cells [27–32]. Dielectrophoresis is a translational motion of neutral mass caused by polarization effects in non-uniform electric field [33,34]. Cells attract small dielectric particles which remain attached to their surface. Particles with high permittivity

($BaTiO_3$, $SrTiO_3$, $NaNbO_3$, and polymer denoted as DP1A) and with low permitivity ($BaSO_4$, SiO_2) were used and the number of particles attached to the cells (e.g. to the murine ascites tumor fibroblasts and to the yeast cells – *Saccharomyces cerevisiae*) was evaluated. The greater the permittivity of the particles and the smaller the conductivity of the suspension the greater the number of attached particles. The greatest number of dielectric particles was attached to the cells in the mitotic phase. The dielectrophoretic effects are exerted by the living cells. The dead cells are dielectrophoretically not active. We observed behaviour of dielectric particles in the neighbourhood of leukocyte cells by a microscope. Some particles exerted the Brownian motion (which looked like a bound Brownian motion) in certain distance from a cell. Attaching a particle to the cell was very fast. We observed also that some particles escaped from the vicinity of the cell.

The experimental results correspond to the predictions based on dielectrophoretic forces. Nonetheless, even forces and mechanisms of different origin can be involved in these effects. Streams and turbulences in the suspension as well as the Brownian motion can transport small particles to the cells where they may be trapped by the "static" membrane field or by same short range molecular mechanisms of the cell. Any indirect measurement method is liable to involve many different mechanisms. The direct measurement methods may be difficult as we cannot separate a part of the biological system and we have to measure it as a whole.

The experimental findings represent an evidence of the long range interaction in biological systems and a pathway for a new view based on the coherent endogenous electromagnetic field. Suggestions given by the experimental results connect the endogenous coherent electromagnetic field with proteins in plasma membranes and/or with protein structures inside cells, e.g. with the cytoskeleton which is a fundamnetal part of the cell dynamic organization. Very low power, a tesseral interference structure, and strong inhomogeneities of the field in the submicrometer region are serious obstacles to measurements. We believe that new measurement methods should be developed. The measurement of the endogenous field is crucial and sets a challenge for research in the near future. It may elucidate the role of coherent vibrations in biological systems and in cancerous cell transformation. Biological systems are very complicated and experiments should exclude possible alternative explanations too.

We may conclude that the long range interactions leading to cooperation and biological order have not been satisfactorily explained yet. Coherence based on energy condensation in vibration states and generation of the endogenous electromagnetic field is more than a plausible mechanism in this respect.

2. Thermodynamic Aspects of Order

2.1 Closed Systems

First of all our concern will be focused to the thermodynamic laws of closed (isolated) systems. In a closed system one form of energy can be transformed into another form but the sum remains constant. Heat is a form of energy and each body possesses a given heat content which is equal to the product of the temperature by a quantity proportional to the mass (called the heat capacity). Heat is a manifestation of the kinetic energy of random motion of the atoms and molecules in the system. Disorganization is the esssential property of that part of energy known as heat. At least a part of the internal energy of a closed system exists in the form of heat. The difference between this energy and the kinetic energy of a macroscopic body (e.g. such as a billiard ball) is in the degree of organization.

The principle of conservation of energy – The First Law of Thermodynamics – is given in the form

$$\mathrm{d}U = \delta Q + \delta W \tag{2.1}$$

where δQ is the quantity of heat supplied to a system, δW is the energy supplied from sources which are not of the "heat" type (e.g. the energy supplied by mechanical work), and $\mathrm{d}U$ is the increase of internal energy.

Heat represents the random part of the internal energy, i.e. the energy distributed in disorganized motion. The function that characterizes the random property of heat is entropy. Entropy is defined by the relation

$$\mathrm{d}\mathcal{S} = \frac{\mathrm{d}Q}{T} \tag{2.2}$$

where T is the temperature in K and $\mathrm{d}Q$ is the quantity of heat supplied in a reversible process.

We may use the Einstein approximation to evaluate the heat content of a system with N atoms at a constant temperature. For the sake of simplicity we assume that the system may be described by $3N$ harmonic oscillators with the same frequency ν. The oscillations of the atoms are independent. The probability of occupation of the energy level $nh\nu$ is proportional to $\exp\left(-nh\nu/kT\right)$

(where h is Planck's constant, k is the Boltzmann constant, and n is an integer). The energy of oscillations in the system is given by the relation

$$U_Q = Q = 3N \frac{\sum_i nh\nu \exp\left(-\frac{nh\nu}{kT}\right)}{\sum_i \exp\left(-\frac{nh\nu}{kT}\right)} = 3N \frac{h\nu}{\exp\left(\frac{h\nu}{kT}\right) - 1} \tag{2.3}$$

where U_Q in Eq. (2.3) is a part of the internal energy caused by random oscillations in the system. The temperature increase $\mathrm{d}T$ results in the increase of the internal energy $\mathrm{d}U_Q$. We may write

$$\mathrm{d}U_Q = \mathrm{d}Q = \frac{\mathrm{d}Q}{\mathrm{d}T}\mathrm{d}T \tag{2.4}$$

(in thermodynamic equilibrium $\mathrm{d}U = 0$ and $\mathrm{d}Q = T\mathrm{d}\mathcal{S} = 0$) and from Eqs (2.2) and (2.3) we obtain the increment of entropy in the form

$$\mathrm{d}\mathcal{S} = \frac{1}{T}\frac{h\nu}{kT^2}\frac{3Nh\nu \exp\left(\frac{h\nu}{kT}\right)}{\left[\exp\left(\frac{h\nu}{kT}\right) - 1\right]^2}\mathrm{d}T\,. \tag{2.5}$$

This simplified example shows the connection of the entropy with the kinetic energy of atoms in the system.

We may assume that at the initial time $t = 0$ the occupation number M of certain energy level $n_0 h\nu$ is greater than the Boltzmann factor, i.e.

$$M > \exp\left(-\frac{n_0 h\nu}{kT}\right).$$

The excess energy $[M - \exp(-n_0 h\nu/kT)]\, h\nu$ relaxes in time $t > 0$ and is distributed among all energy levels in the system. Under the steady state conditions (in thermodynamic equilibrium) the temperature of the system is $T + \Delta T$ and entropy of the system increased.

Boltzmann related entropy of a system in a given state to the probability of finding the system in this state. Entropy is an additive quantity. The entropy $\mathcal{S}$ of several independent systems is given by the sum of their entropies $\mathcal{S}_i$. The total probability of finding all the systems (which are mutually independent) in a given state is equal to the product of the probabilities P_i of the separate systems. Thus

$$\mathcal{S} = \sum_i \mathcal{S}_i\,, \quad P = \prod^i P_i\,. \tag{2.6}$$

Boltzmann postulated validity of the relation

$$S = k \ln P_i \,. \tag{2.7}$$

The thermodynamic probability is related to micro distributions in the system. The number of microstates corresponding to a given macrostate is a measure of the probability of occurrence of the thermodynamic state.

Experience shows that it is imposible to extract energy from a system with disorganised random motion without other energy supply (i.e. to gain energy by cooling the system). Therefore, we cannot decrease entropy with no energy supply and obtain useful energy. This is expressed in The Second Law of Thermodynamics.

Theory of probability enable us to assess the possibility of spontaneous violation of The Second Law of Thermodynamics in a disorganized system. We accept an idealized model of a system composed of N oscillators whose energy levels correspond to the thermodynamic equilibrium distribution. This system is connected with another system which may accept or supply energy quanta but only as a passive "participant". Each oscillator may change its energy: the energy may increase or decrease by one quantum $h\nu$ or remain unchanged. Therefore, three possibilities for each oscillator exist (the occupation numbers are $n-1$, n, $n+1$ where $n > 0$ is the integer determining the energy level of each oscillator) and we assume that the energy levels have the same "a priori" probabilities. (The lowest energy level of an oscillator is $\frac{1}{2}h\nu$; therefore the energy levels are $\frac{1}{2}h\nu + nh\nu$. We do not take the lowest energy level into consideration.) The whole system of N oscillators has $3^N \approx 10^{N/2}$ combinations of these three energy states (and only one of them is the desired cooling effect). The whole system of oscillators may be cooled if each oscillator looses energy $h\nu$. If annihilation or creation of one quantum of energy $h\nu$ in an oscillator may be realized within the time $10^{-10}\,$s (but no simultaneous transitions occur) then all the 3^N combinations may be realized within a time period

$$\Delta t \doteq 10^{N/2} \times 10^{-10}\,\mathrm{s} = 10^{(N/2)-10}\,\mathrm{s} \,. \tag{2.8}$$

For a microscopic system with $N = 2000$ oscillators the time period $\Delta t = 10^{990}\,$s which is about 10^{982} years. (We note that in this time period some combinations can appear several times and others need not occur at all.) As any macroscopic system contains a very large amount of entities (about 2.7×10^{19} molecules in $1\,\mathrm{cm}^3$ of a gas) the probability of spontaneous cooling the system is assumed to be out of the question. Similar result may be derived for "self-arrangement" of the system.

Reversible energy processes and reversible energy transformations in a closed isolated system leave the entropy unchanged. Irreversibility of any process leads to the increase of the entropy. All irreversible changes in a closed system lead to increase of the entropy until the entropy reaches the maximum value. The maximum value of entropy is a condition for the thermodynamic equilibrium state. All processes in a closed system satisfy, therefore, the relation

$$\mathcal{S}_\mathrm{o} \geq \mathcal{S}_\mathrm{b} \,, \quad \mathrm{d}\mathcal{S} \geq 0 \tag{2.9}$$

where $\mathcal{S}_\mathrm{b}$ is the entropy before and $\mathcal{S}_\mathrm{o}$ after the process in question. Inside a closed system in thermodynamic equilibrium organization is hardly possible. Such system cannot contain information.

The internal energy U of a system may be greater than its heat content ($\int T\mathrm{d}S$). In an isothermal-isometric system the part of energy over TS, i.e.

$$F = U - TS \tag{2.10}$$

is called the free energy. Any irreversible change in the system leads to decrease of the free energy. The equilibrium state has a minimum of the free energy.

The Gibbs potential has a similar meaning as the free energy but in an isothermal-isobaric system. The Gibbs potential is given by the relation

$$G = U - TS + pV = F + pV \tag{2.11}$$

where p is the pressure and V is the volume of the system. Description of the thermodynamic potentials may be found e.g. in [35].

As all real processes in nature are irreversible the free energy as well as the Gibbs potential tend to their minimum values and entropy to its maximum value. Nature favours the maximum entropy in the closed systems. We may ask a question whether Nature sides with disorganization.

2.2 Open Systems

We discussed the thermodynamics of closed systems with the conclusion that closed systems in thermodynamic equilibrium reach the maximum value of the entropy which means disorganization. From the point of view of thermodynamics any kind of organization depends on the difference between the internal energy U and its heat part determining entropy. For instance, in an isothermal-isometric system organization may depend on the free energy $F = U - TS$ of the system. We will discuss a simple example – a system with different energy levels. In the equilibrium state the probability that the system has the energy level E_n depends on the Boltzmann factor $\exp(-E_n/kT)$. At a very low temperature only the low energy levels are occupied. The term TS in the relation for the free energy may be small and creation of organized structures is possible. Such organized structures may be e.g. of the crystalline or phase transition types.

The open systems are not isolated from their surroundings. The open systems exchange matter and energy with the ambient medium. The change of entropy $\mathrm{d}\mathcal{S}$ in time $\mathrm{d}t$ may be decomposed into two parts [4,36]

$$\mathrm{d}\mathcal{S} = \mathrm{d}\mathcal{S}_\mathrm{f} + \mathrm{d}\mathcal{S}_\mathrm{s} \tag{2.12}$$

where $\mathrm{d}\mathcal{S}_\mathrm{f}$ is the entropy that flows to or from the ambient medium and $\mathrm{d}\mathcal{S}_\mathrm{s}$ is the entropy produced in the open system. The production of entropy is caused by irreversible processes such as e.g. diffusion or collisions. The Second Law of Thermodynamics yields

$$\mathrm{d}\mathcal{S}_\mathrm{s} \geq 0 \tag{2.13}$$

where $\mathrm{d}\mathcal{S}_\mathrm{s} = 0$ in thermodynamic equilibrium. In a closed system the flow of the entropy outwards from (or inwards into) the system does not exist ($\mathrm{d}\mathcal{S}_\mathrm{f} = 0$). One gets from Eqs (2.12) and (2.13)

$$\mathrm{d}\mathcal{S} = \mathrm{d}\mathcal{S}_\mathrm{s} \geq 0\ . \tag{2.14}$$

In an open system the change of the entropy $\mathrm{d}\mathcal{S}_\mathrm{f}$ by flow may be positive

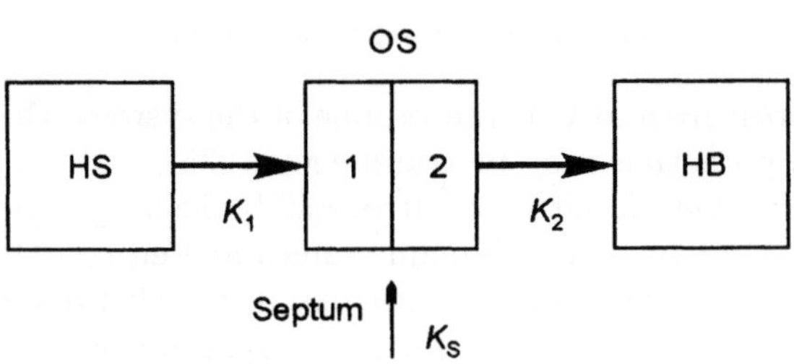

Fig. 2.1 A schematic picture of the open system (OS) with the source of energy (the source of heat – HS), and the heat bath (HB). The flow of energy is indicated by the thick horizontal line segments with arrows.

(inwards into the open system) as well as negative (outwards from the open system). For a negative flow the total entropy of the open system may decrease. In the evolution the system in the final state may have lower entropy than the system in the initial state. If $\mathrm{d}\mathcal{S}_\mathrm{f} < 0$ and $|\mathrm{d}\mathcal{S}_\mathrm{f}| > |\mathrm{d}\mathcal{S}_\mathrm{s}|$ one gets from Eq. (2.12)

$$\mathcal{S}_2 - \mathcal{S}_1 = \int_1^2 \mathrm{d}\mathcal{S} < 0\ . \tag{2.15}$$

In the stationary state of the open system the total change of the entropy equals zero, i.e. $\mathrm{d}\mathcal{S} = 0$ and from Eq. (2.12)

$$\mathrm{d}\mathcal{S}_\mathrm{f} = -\mathrm{d}\mathcal{S}_\mathrm{s} \leq 0\ . \tag{2.16}$$

If the open system has a sufficiently high negative flow of the entropy (i.e. the flow of the entropy from the system to the ambient medium) then the random part of the internal energy may be small in comparison with the total internal energy. Some organization can occur. The flow of entropy outwards from the system decreases randomness.

If energy is channelled from a source through a system to a heat bath a non-equilibrium state is created. We accept a model of an open system with energy supply from the source of heat (Fig. 2.1). The open system OS is composed of two equivalent parts (i.e. of the part 1 and of the part 2) separated by a septum of a finite thermal conductivity K_S. The thermal conductivities of the part 1 as well as of the part 2 of the system are so high that the temperature gradients may be neglected. The source of heat HS with the temperature T_S supplies heat to the part 1 of the system (with the temperature T_1). The source is large enough and no temperature change occurs. The quantity of the supplied heat depends on the thermal conductivity K_1 between the source of heat and the part 1 of the system. The heat supplied is channelled through the open system to the heat bath HB (with the temperature T_C). The thermal conductivity $K_2 \to \infty$ (the temperature T_2 of the part 2 is, therefore, equal to T_C). For the temperatures T_S, T_1, T_2, and T_C one can write

$$T_S > T_1 > T_2 = T_C \; . \tag{2.17}$$

There is a temperature difference between the part 1 and the part 2 of the open system.

Before we connected the source of heat HS to the open system OS (Fig. 2.1) this system was in thermodynamic equilibrium with the heat bath, i.e.

$$T_1 = T_2 = T_C \; ,$$

and the total entropy given by the sum of the entropy $\mathcal{S}_1$ and $\mathcal{S}_2$ of the part 1 and of the part 2, respectively, was given by

$$\mathcal{S} = \mathcal{S}_1 + \mathcal{S}_2 = 2\mathcal{S}_1 \; . \tag{2.18}$$

After connection of the source of heat (HS) to the open system heat is supplied to the system and channelled to the heat bath. The infinitesimal amount of heat supplied to the part 1 of the system is given by $\mathrm{d}Q = m_1 c_v \mathrm{d}T$ where c_v is the specific heat at constant volume, and m_1 is the mass of the part 1. After reaching the steady state conditions the temperature of the part 1 is T_1 and the total amount of the supplied heat is given by

$$\Delta Q = m_1 c_v \left(T_1 - T_C\right) \tag{2.19}$$

which increases the internal energy ($\Delta Q = \Delta U$). The entropy of the part 2 remains unchanged ($\mathcal{S}_2 = \mathcal{S}_1$) and the entropy of the part 1 is

$$\mathcal{S}_1' = \mathcal{S}_1 + m_1 c_v \int_{T_C}^{T_1} \frac{\mathrm{d}T}{T} = \mathcal{S}_1 + m_1 c_v \ln\left(\frac{T_1}{T_C}\right) . \tag{2.20}$$

The total entropy of the system is

$$\mathcal{S}' = 2\mathcal{S}_1 + m_1 c_v \ln\left(\frac{T_1}{T_C}\right) . \tag{2.21}$$

If we separate the system in question from the source of heat HS as well as from the heat bath HB the thermodynamic equilibrium state is reached when $T_1'' = T_2''$. As both the parts of the system are identical the temperature $T_1'' = (T_1 + T_C)/2$ and the total entropy is given by

$$\mathcal{S}'' = 2\mathcal{S}_1 + 2m_1 c_v \ln\left(\frac{T_1 + T_C}{2T_C}\right) . \tag{2.22}$$

For $T_1 > T_C$ the entropy of the whole system $\mathcal{S}' < \mathcal{S}''$. The flow of energy through the system forms the free energy and in this sense causes certain type of "order".

We note that the system in question is linear and its properties are not altered by the energy flow. Probability of any event remains unchanged. The Second Law of Thermodynamics is a probability law. As far as the probability is not altered we cannot expect any special effect.

2.3 Selforganization

Energy flow from a source of heat through an open system to a heat bath can shift the open system from thermodynamic equilibrium. The system remains in the steady state as long as heat transfer continues. The energy flow across the system is very likely a necessary condition for certain type of organization. Nevertheless, the energy flow by itself has not any special consequence such as selforganization.

We will further discuss the concept of entropy given by the relation (Appendix 13.9 [37])

$$\mathcal{S} = kN \sum_i P_i \ln \frac{1}{P_i} \tag{2.23}$$

where

$$P_i = \lim_{N \to \infty} \frac{N_i}{N}$$

is the probability of a thermodynamic state (N_i is the number of entities in the i-th cell, and $N = \sum_i N_i$). Therefore, Eq. (2.23) may be written as [37]

$$\mathcal{S} = kN \sum_i \frac{N_i}{N} \ln \frac{N}{N_i} = k \left[N \ln N - \sum_i N_i \ln N_i \right] . \tag{2.24}$$

The differentiation of $\mathcal{S}$ with respect to time determines the change of entropy in the unit time

$$\frac{\partial \mathcal{S}}{\partial t} = -k \sum_i \left(\frac{\partial N_i}{\partial t} \ln N_i + \frac{\partial N_i}{\partial t} \right) = -k \sum_i \frac{\partial N_i}{\partial t} \ln N_i \tag{2.25}$$

as

$$\sum_i \frac{\partial N_i}{\partial t} = \frac{\partial N}{\partial t} = 0 .$$

The derivative $\partial N_i / \partial t$ in Eq. (2.25) depends on the probability of the transfer of entities (particles) from the i-th cell to the other cells (that are denoted as the j-th cells) and vice versa, i.e. on the transfer probabilities α_{ij} and α_{ji}, respectively. The number of particles transferred in the unit time between the i-th cell and the j-th cells is given by

$$\frac{\partial N_i}{\partial t} = - \sum_j \alpha_{ij} N_i + \sum_j \alpha_{ji} N_j . \tag{2.26}$$

From Eq. (2.25) one gets relation

$$\frac{\partial \mathcal{S}}{\partial t} = k \sum_i \sum_j (\alpha_{ij} N_i - \alpha_{ji} N_j) \ln N_i \tag{2.27}$$

which determines the change of entropy. Under usual conditions the transfer probabilities are assumed to be equal, i.e. $\alpha_{ij} = \alpha_{ji}$ and a maximum of the entropy is reached for $N_i = N_j$ for all i and j. In a general case $\alpha_{ij} \neq \alpha_{ji}$ and the equality $N_i = N_j$ is not satisfied under the steady state conditions.

We will discuss the possibility of the change from the state $\alpha_{ij} = \alpha_{ji}$ to the state $\alpha_{ij} \neq \alpha_{ji}$. In a non-linear system a small disturbance, such as a slight change of initial conditions or of the energy supply, can result in a big difference in behaviour of the system at a later time [38]. As a consequence, properties of non-linear systems depend on excitation and even the probabilities α_{ij} and α_{ji} may be subject to alterations caused by non-linear effects. Nevertheless, in a non-linear system Eq. (2.24) deserves deeper analysis on account of the breakdown of the superposition principle.

As an example we may analyze a simple hypothetical model system containing two cells and a finite number N of entities (particles). From Eq. (2.27) one gets

$$\frac{\partial S}{\partial t} = k\,(\alpha_{12}N_1 - \alpha_{21}N_2)\,(\ln N_1 - \ln N_2)\ . \tag{2.28}$$

First of all we will assume that the sytem is isolated from the surroundings. For $\alpha_{12} = \alpha_{21}$ the equilibrium conditions are reached at $N_1 = N_2 = N/2$ (we neglect fluctuations). The time evolution of the entropy of the system from the state $N_1 = 1$, $N_2 = N - 1$ to equilibrium is given in Fig. 2.2 (the full curve denoted by a). At the time t_{ch} we connect the system to the source of energy and to the heat bath (as is shown in Fig. 2.1). We assume that due to the energy flow through the system the probabilities α_{12} and α_{21} are changed so that $\alpha_{12} \neq \alpha_{21}$. The evolution of the system from equilibrium where $N_1 = N_2$ to a new steady state is shown in Fig. 2.2 (the dashed curve denoted by b). The entropy decreased as a result of the energy flow through the system.

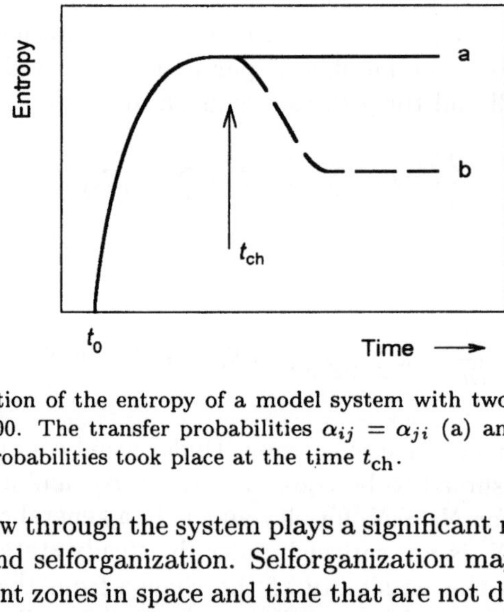

Fig. 2.2 Time evolution of the entropy of a model system with two cells and the number of particles $N = 2000$. The transfer probabilities $\alpha_{ij} = \alpha_{ji}$ (a) and $\alpha_{ij} \neq \alpha_{ji}$ (b). The change of transfer probabilities took place at the time t_{ch}.

The energy flow through the system plays a significant role in the processes of organization and selforganization. Selforganization may be understood as creation of coherent zones in space and time that are not damaged by disturbances of random nature from the surroundings. (Fundamental information on selforganization may be found e.g. in [4,39–43]. We will not discuss here the general principles of determinism, randomness, and chaos; they may be found e.g. in [38,44].) The energy of the coherent zones has to be greater than the energy in thermodynamic equilibrium. Interactions (in the most general sense) have to exist to keep order in the coherent zones. We may assume that the interactions among polar particles and polar systems are important to maintain coherence. The electromagnetic forces acting among oscillating parts of the systems are not screened by charges with low mobility (if the oscillation frequencies are high, Appendix 13.13) and may have a fundamental function in the biological systems.

It is generally accepted that selforganization is conditioned by non-linear properties of open systems. Non-linearity may be the inner property of the open system but the non-linear interactions of then open system with the heat bath may exist as well. Events which are independent in linear systems may be coupled in non-linear systems. The vibration modes in non-linear systems are not mutually independent and the non-linear mechanism transfer energy among them. The total probability of finding a non-linear system in certain thermodynamic state is not equal to the product of probabilities of its separate parts. Relations based on the probability of independent events cannot be used.

Interactions and energy transfer in non-linear systems have special features. In oscillating non-linear systems some spectral components are preferred which can result in energy condensation. The energy of some spectral components may be diminished while the energy of the other spectral components may increase by excitation and their frequencies may be shifted. Energy condensation is a precursor of organization. Bifurcation is another essential feature which may appear in non-linear systems. The processes in non-linear systems may proceed along various branches of time evolution. From the point of view of thermodynamics biological systems are non-linear open systems far from thermodynamic equilibrium.

3. Normal Modes

3.1 The Vibration Modes of a String

Before we start an analysis of vibrations in discrete systems of particles we shall deal with the vibration modes of a continuous medium to point out its fundamental properties. We will consider vibrations of a one-dimensional continuous string of the length ℓ. At the point x (where x is the space coordinate along the string) and at the time t the displacement of the string is $q(x,t)$. Vibration of a string is described by the one-dimensional wave equation

$$\frac{\partial^2 q(x,t)}{\partial x^2} = \frac{1}{c^2}\frac{\partial^2 q(x,t)}{\partial t^2} \tag{3.1}$$

where c is the velocity of propagation of elastic waves. Assuming the end points of the string to be fixed and both directions of propagation to be equivalent we may find the solution of Eq. (3.1) in the form of standing waves

$$q(x,t) = A\sin(k_j x)\cos(\omega_j t) \tag{3.2}$$

where k_j is the absolute value of the wave vector (which is often called the wave constant or the propagation constant), ω_j is the angular frequency, and A is the amplitude. As the wavelength

$$\lambda_j = \frac{2\ell}{j} \quad \text{for} \quad j = 1,\, 2,\, 3 \ldots \tag{3.3}$$

the wave constant and the frequency of vibrations are given by

$$k_j = \frac{2\pi}{\lambda_j} = \frac{\pi}{\ell}j\,, \qquad \omega_j = \frac{2\pi c}{\lambda_j} = \frac{\pi c}{\ell}j\,. \tag{3.4}$$

The frequency spectrum of the vibrations of a continuous string is discrete. One frequency corresponds to each value of the wave constant k_j, i.e. to each integer value j. The frequency is a linear function of the wave constant k_j. There is no upper limit of the frequency. The component of vibrations related to the wave constant k_j is called the vibration mode. The total energy of vibrations is composed of energy of vibration modes. We have to emphasize here that energy of a vibration mode is delocalized along the string.

3.2 Vibrations in an Infinite One-Dimensional Chain of Identical Particles

We will discuss vibrations in an ordered array of particles (atoms). We will use a symmetric lattice to find the possible modes of vibrations [45–47]. We assume one-dimensional chain of equidistant identical particles with the equilibrium positions along a straight line of infinite length. The distance between the particles is a. Dimensions of the particles are much smaller than the distance a between them. The particles may be displaced from their equilibrium position as is shown in Fig. 3.1.

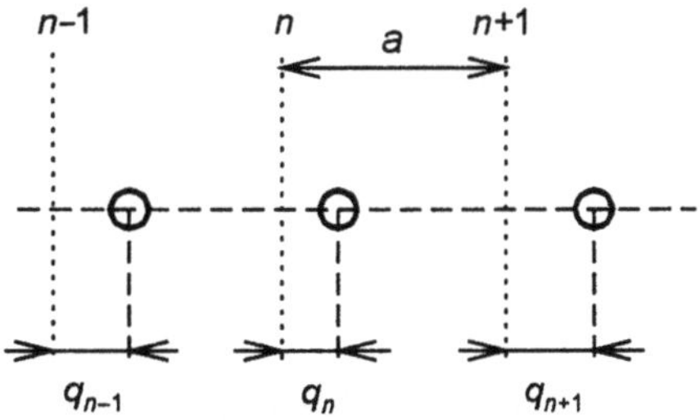

Fig. 3.1 One-dimensional chain of identical particles. The points of intersections of the dotted vertical line segments with the dashed horizontal line represent the equilibrium positions and the circles denote the displaced particles.

We will assume that Hook's law (ut tensio, sic vis) is obeyed and that the interaction between particles exists between the nearest neighbours. A force acting upon a particle to restore the equilibrium position is proportional to the displacement, i.e. the linear dependence fq_n is valid (where q_n is the displacement of the n-th atom and f is the elastic force constant) for the restoring elastic force. The equation of motion is given by

$$m\ddot{q}_n = -f(q_n - q_{n-1}) - f(q_n - q_{n+1}) = f(q_{n-1} + q_{n+1} - 2q_n) \qquad (3.5)$$

where $\ddot{q}_n$ is the second derivative of q_n with respect to time and m is the mass of the particle. We assume solution of Eq. (3.5) in the form of propagating waves

$$q_n = \exp\left[-\mathrm{i}(\omega t - kna)\right], \qquad q_{n-1} = \exp\left\{-\mathrm{i}\left[\omega t - k(n-1)a\right]\right\},$$
$$q_{n+1} = \exp\left\{-\mathrm{i}\left[\omega t - k(n+1)a\right]\right\} \qquad (3.6)$$

where ω is the angular frequency of vibrations, $k = \omega/c = 2\pi/\lambda$ (c is the velocity of propagation and λ is the wavelength) is the wave constant (generally – the wave vector), and n is an integer relating the position of a particle (atom, molecule, etc.) to the origin. Vibrations along the chain are described by a wave which may be expanded into the Fourier series containing terms given by Eq. (3.6) with different frequencies. (The Fourier expansion can be used provided that the system in question is linear and the principle of superposition is valid.)

After substitution of Eq. (3.6) into Eq. (3.5) we get

$$m\omega^2 = -f\left[\exp(-ika) + \exp(ika) - 2\right] = -2f\left[\cos(ka) - 1\right] = 4f\sin^2\left(\frac{ka}{2}\right). \tag{3.7}$$

Therefore, the angular frequency of vibrations is given by the dispersion relation

$$\omega = \omega_{\max}\left|\sin\left(\frac{ka}{2}\right)\right|, \qquad \omega_{\max} = 2\sqrt{\frac{f}{m}} \tag{3.8}$$

where $\omega_{\max}$ is the maximum angular frequency (i.e. the upper limit of fre-

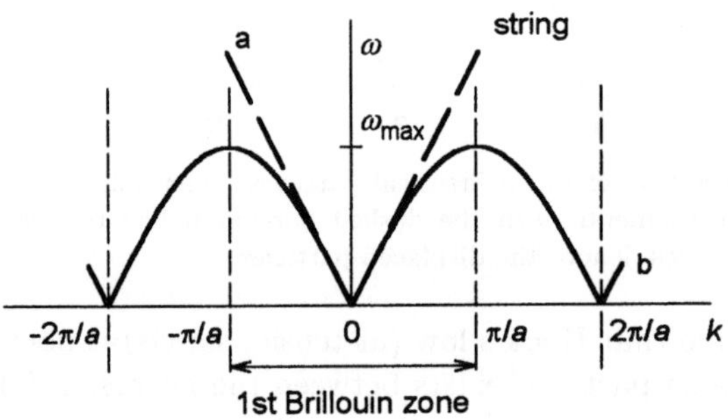

Fig. 3.2 Frequency of vibrations of a continuous string (the dashed lines – a) and of a one-dimensional chain of identical particles (the full line – b) as a function of the wave constant k. The first Brillouin zone is indicated.

quency exists which is a difference in contrast with a continuous string). A mode with the frequency higher than $\omega_{\max}$ cannot propagate along the chain. The angular frequency ω as a function of k is shown in Fig. 3.2. The frequency ω is a periodic function of the wave constant k as follows from Eq. (3.8). The same value of the frequency is determined for the wave constant k and $k + 2\pi g/a$ where g is an integer ($g = \pm 1, \pm 2, \pm 3, \ldots$). The state of vibration of the array of particles corresponding to the wave constant k is the same as the state for a wave constant $k + 2\pi g/a$, i.e. the solutions are identical. A unique relationship between the state of vibrations of the chain and the wave constant k exists for k values confined to a range of values $2\pi/a$. Usually the range is chosen in the following way

$$-\frac{\pi}{a} \leq k < \frac{\pi}{a}. \tag{3.9}$$

The region of the k values given by Eq. (3.9) is referred to as the first Brillouin zone.

The velocity of propagation of the waves depends on the frequency, i.e. $c = \omega/k$. For the high frequencies near the boundary value $\omega_{\max}$ (i.e. for the short wavelengths) the velocity of propagation is smaller than for the low frequencies (i.e. for the long wavelengths). In the continuous approximation the frequency is proportional to the wave constant (Fig. 3.2) and the velocity of propagation is independent of the wavelength.

We have to mention here that the analysis given above is highly simplified and holds only for identical particles in the one-dimensional infinite lattice with the translation symmetry. The atoms in a real molecule are not equidistant and identical – they may have different masses and charges. The force constant f is not the same for all atoms, the potential valley may be asymmetric, interactions need not be confined to the closest neighbours, and the chain of atoms is finite and folded. A similar rationale holds for "particles" in a polymer structure.

3.3 The Number of Vibration Modes

Infinite chains dealt with in the preceding section do not correspond to real systems. We want to enumerate the possible modes of vibrations and we accept, therefore, finite lattice and introduce the boundary conditions into calculations [46]. The space confinement is the only difference with respect to the preceding section. For the sake of simlicity we will assume $(N+1)$ identical particles. The particles at both ends of the chain cannot move (i.e. the zero-th and the N-th particle). The remaining $(N-1)$ particles undergo vibrations. The general solution of Eq. (3.5) may be given for each wavelength (for each wave constant) as a sum of waves propagating in opposite directions: one propagating to the right, the other to the left. The displacement of the n-th particle is given by

$$q_n(t) = \{A_1 \exp[\mathrm{i}(kna + \beta_1)] + A_2 \exp[-\mathrm{i}(kna - \beta_2)]\} \exp(-\mathrm{i}\omega t) \tag{3.10}$$

where A_1 and A_2 are the amplitudes, and β_1 and β_2 are the phase angles. The boundary conditions are given by the relations

$$q_0(t) = 0 , \qquad q_N(t) = 0 . \tag{3.11}$$

Using the first boundary condition given by Eq. (3.11) we get from Eq. (3.10) that $A_1 = -A_2$ and $\beta_1 = \beta_2$ (we will choose $\beta_1 = \beta_2 = 0$). From Eq. (3.10) we get

$$q_n(t) = 2A_1 \sin(kna)\,[\sin(\omega t) + \mathrm{i}\cos(\omega t)] \tag{3.12}$$

which represents a standing wave. Using the second boundary condition $q_N(t) = 0$ given in Eq. (3.11) we get from the Eq. (3.12) that

$$\sin(kNa) = 0 \quad \text{which is satisfied for} \quad k = \frac{\pi}{Na}r \tag{3.13}$$

where r is an integer. In this case k is limited to positive values from 0 to π/a. For $r = N$ we get that $k = \pi/a$. For $r = 0$ then $k = 0$. In both these cases the particles at the ends of the chain are at rest as follows from Eq. (3.12). The wave constant may have, therefore, $N - 1$ values with $r = 1, 2, 3, \ldots, N - 1$. The number of vibration modes is given by the number of wave constants k which corresponds to the number of mobile particles.

A similar result may be derived using the cyclic (or periodic) boundary conditions introduced by Born and von Karman. We may assume that $n' = n + N$ where N is a large number ($N \gg 1$). If N satisfy the condition

$$q_n(t) = q_{n+N}(t) \tag{3.14}$$

i.e.

$$\exp(ikn'a) = \exp[ik(n+N)a] = \exp(ikna)$$

then $kNa = 2\pi g$ where g is an integer. The wave constant is given by $k = 2\pi g/Na$. As k (in the first Brillouin zone) is confined by the condition given by Eq. (3.9) the possible values of g are given by the relation $-N/2 \leq g < N/2$.

3.4 Equivalence of a Vibration Mode and a Harmonic Oscillator

Equivalence of a vibration mode and a harmonic oscillator is proved e.g. in [46]. A harmonic oscillator is described by the relation $M\ddot{y} = -fy$ which for $y = \sin(\omega t)$ yields $\omega^2 = f/M$ (where M is the mass). The total energy E of the harmonic oscillator is given by the sum of the kinetic energy T and of the potential energy U.

$$E = \frac{p^2}{2M} + \frac{1}{2}M\omega^2 y^2 = \frac{1}{2}M\omega^2\left[\cos^2(\omega t) + \sin^2(\omega t)\right] = \frac{1}{2}M\omega^2 \tag{3.15}$$

where the momentum $p = M\mathrm{d}y/\mathrm{d}t$.

We assume a vibration mode corresponding to a standing wave with certain value of the wave constant k as is given by Eq. (3.12). The kinetic energy is given by the relation

$$T_k = \frac{1}{2m}\sum_n p_n^2 = \frac{1}{2}m\omega^2\cos^2(\omega t)\sum_n \sin^2(kna) \tag{3.16}$$

where $p_n = m\mathrm{d}q_n(t)/\mathrm{d}t$ and $q_n(t) = \sin(kna)\sin(\omega t)$ (the real part in Eq. (3.12) for $2A_1 = 1$). The potential energy of all particles in the chain due to the vibration mode k is a function of all coordinates q_n, i.e.

$$U_k = U_k(q_0, q_1, \ldots, q_n, \ldots, q_N) .$$

The force acting upon a particle positioned at na with displacement q_n is therefore

$$-\frac{\partial U_k}{\partial q_n} = m\frac{\mathrm{d}^2 q_n}{\mathrm{d}t^2} = f(q_{n-1} + q_{n+1} - 2q_n) \tag{3.17}$$

where Eq. (3.5) was used. The total potential energy is given by the relation

$$U_k = \frac{1}{2} f \sum_n \left(2q_n^2 - q_n q_{n-1} - q_n q_{n+1}\right) \tag{3.18}$$

(each of the mixed terms appears in the sum twice; therefore, we have to divide the sum by 2). Replacing q_n, q_{n-1}, and q_{n+1} by $\sin(kna)\sin(\omega t)$, $\sin[k(n-1)a]\sin(\omega t)$, and $\sin[k(n+1)a]\sin(\omega t)$, respectively, we get after some elementary arrangements and manipulations that

$$\begin{aligned} U_k &= f\sin^2(\omega t)\left[1 - \cos^2\left(\frac{ka}{2}\right) + \sin^2\left(\frac{ka}{2}\right)\right]\sum_n \sin^2(kna) \\ &= 2f\sin^2\left(\frac{ka}{2}\right)\sin^2(\omega t)\sum_n \sin^2(kna) . \end{aligned} \tag{3.19}$$

We make use of the relation $2f\sin^2(ka/2) = \frac{1}{2}m\omega^2$ following from Eq. (3.7). We get from Eq. (3.19)

$$U_k = \frac{1}{2}Bm\omega^2\sin^2(\omega t) \qquad \text{where} \qquad B = \sum_n \sin^2(kna) . \tag{3.20}$$

From Eqs (3.16) and (3.19) we get for the total energy of a vibration mode the relation

$$E = \frac{1}{2}Bm\omega^2\left[\cos^2(\omega t) + \sin^2(\omega t)\right] = \frac{1}{2}Bm\omega^2 \tag{3.21}$$

which has the same form as Eq. (3.15) for a harmonic oscillator. The equivalence of a vibration mode and a harmonic oscillator is proved.

The average energy of a vibration mode with angular frequency ω in thermodynamic equilibrium is given by Planck's relation

$$E = \frac{\hbar\omega}{\exp\left(\dfrac{\hbar\omega}{kT}\right) - 1}$$

(where k is the Boltzmann constant and $\hbar = h/2\pi$) and the occupation number (the number of quanta of energy) in the vibration mode is given by

$$n = \frac{1}{\exp\left(\frac{\hbar\omega}{kT}\right) - 1} .$$

These quanta of energy are commonly reffered to as phonons. A phonon is a quasi-particle with the energy $\hbar\omega$ and the momentum

$$p = \frac{\hbar\omega}{c_k}$$

where c_k is the velocity of propagation of the vibration mode.

3.5 Vibrations in One-Dimensional Chain with Particles of Two Types

We will consider a one-dimensional chain containing two types of particles as is illustrated in Fig. 3.3 [46]. The distance between two neighbouring particles is a. We assume that only the nearest particles interact. The equations of motion are analogous with Eq. (3.5)

$$\begin{aligned} M\ddot{q}_{2n} &= f\left(q_{2n-1} + q_{2n+1} - 2q_{2n}\right) \\ m\ddot{q}_{2n+1} &= f\left(q_{2n} + q_{2n+2} - 2q_{2n+1}\right) \end{aligned} \tag{3.22}$$

where M, m are masses of the particles. The numbering of particles is done in such a way that odd numbers correspond to mass m and even numbers correspond to mass M.

We will solve Eq. (3.22) by propagating waves

$$\begin{aligned} q_{2n} &= A\exp\left[-\mathrm{i}(\omega t - k2na)\right] \\ q_{2n+1} &= B\exp\left\{-\mathrm{i}\left[(\omega t - k(2n+1)a]\right)\right\} \end{aligned} \tag{3.23}$$

where A and B are the amplitudes of vibrations corresponding to particles of mass M and m, respectively. After substitution of Eq. (3.23) into (3.22) we get

$$\begin{aligned} -MA\omega^2 &= f\left[B\exp\left(-\mathrm{i}ka\right) + B\exp\left(\mathrm{i}ka\right) - 2A\right] \\ -mB\omega^2 &= f\left[A\exp\left(-\mathrm{i}ka\right) + A\exp\left(\mathrm{i}ka\right) - 2B\right] \end{aligned} \tag{3.24}$$

and further

$$\begin{aligned} (M\omega^2 - 2f)A + 2Bf\cos(ka) &= 0 \\ (m\omega^2 - 2f)B + 2Af\cos(ka) &= 0 . \end{aligned} \tag{3.25}$$

Nonvanishing solutions for A and B may be found if the determinant of the coefficient at A and B vanishes. Therefore, the determinant

$$\begin{vmatrix} (M\omega^2 - 2f) & , & 2f\cos(ka) \\ 2f\cos(ka) & , & (m\omega^2 - 2f) \end{vmatrix} = 0 \tag{3.26}$$

yields

$$\omega^4 - 2f\omega^2\left(\frac{M+m}{mM}\right) + \frac{4f^2}{mM}\left[1 - \cos^2(ka)\right] = 0\,. \tag{3.27}$$

From Eq. (3.27) one may derive the dispersion relation

$$\omega_\pm^2 = f\left(\frac{1}{m} + \frac{1}{M}\right) \pm f\sqrt{\left(\frac{1}{m} + \frac{1}{M}\right)^2 - \frac{4\sin^2(ka)}{mM}}\,. \tag{3.28}$$

As $\omega_\pm$ has to be positive each value of $\omega_\pm^2$ gives only a single value for $\omega_\pm$.

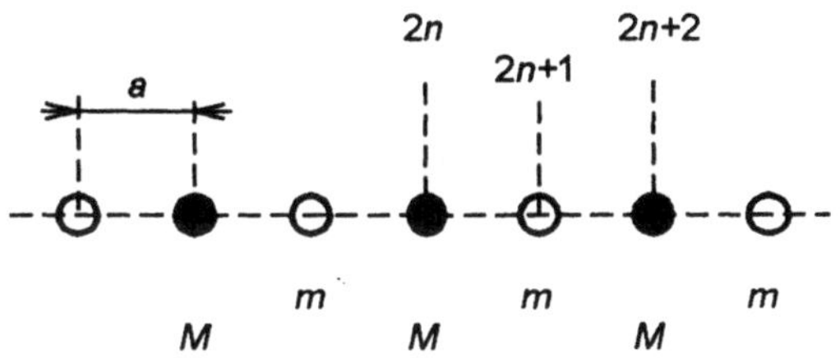

Fig. 3.3 A one-dimensional chain of particles with masses M and m. The particles are equidistant.

The two particle chain has two angular frequencies ω_+ and ω_- corresponding to a single value of the wave constant k. Therefore, the vibrations have two branches of frequencies. The ω_+ branch is called the optical branch of vibrations and the ω_- branch is called the acoustical branch of vibrations. We will assume that $M > m$. We will investigate two limit cases for $\sin(ka) = 0$ and $|\sin(ka)| = 1$. For $k = 0$ ($\sin(ka) = 0$) one obtains

$$\omega_+ = \sqrt{2f\left(\frac{1}{m} + \frac{1}{M}\right)}, \qquad \omega_- = 0\,. \tag{3.29}$$

As follows from Eq. (3.28) frequency is a periodic function of the wave constant k. The first Brillouin zone limits the value of k to the range from $-\pi/2a$ to $+\pi/2a$. For $k = \pm\pi/2a$ ($|\sin(ka)| = 1$) we get

$$\omega_+ = \sqrt{\frac{2f}{m}} \quad \text{and} \quad \omega_- = \sqrt{\frac{2f}{M}}\,. \tag{3.30}$$

The frequency of vibration modes as a function of the wave constant for the acoustical branch of vibrations and for the optical branch of vibrations is visualised in Fig. 3.4. We will point out an important difference between the two branches. We may calculate the ratio of the amplitudes A and B for the optical and acoustical branches. From Eqs (3.25) and (3.29) we may get for $k = 0$

$$\begin{aligned} A &= B \quad \text{for the acoustical branch} \\ MA &= -mB \quad \text{for the optical branch .} \end{aligned} \tag{3.31}$$

In the acoustical branch of vibrations all particles move in the same

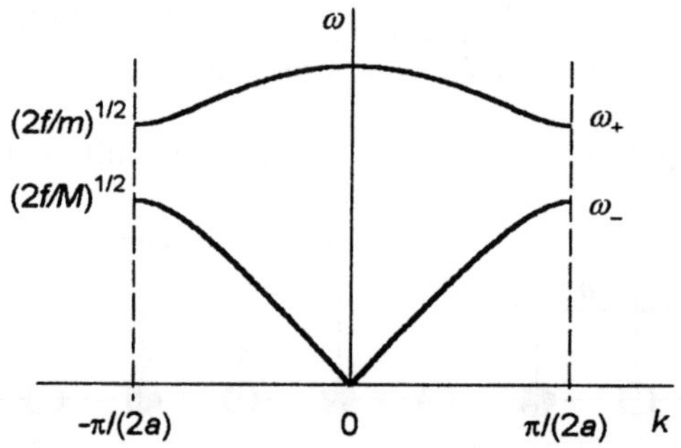

Fig. 3.4 The optical (ω_+) and acoustical (ω_-) branches of vibrations in a chain with two types of particles.

direction. In the optical branch the two types of particles move in opposite directions; the centre of gravity of them remains at rest.

We should remark here that the optical branches of vibrations are responsible for absorption of electromagnetic waves, e.g. in ionic crystals.

3.6 One-Dimensional Disordered Chain

We described the basic properties of one-dimensional ordered systems with translation symmetry in space. In general, biological systems have not symmetric configurations of this type. Space order with "exact" symmetry elements (as may be found in crystalline structures) is lacking. Nevertheless, we can find some exceptions. For instance, protofilaments in the microtubules have certain type of translation symmetry as they are composed from tubulin molecules arranged in an ordered one-dimensionam chain. The actin filaments have similar translation symmetry too.

We will examine systems with configuration disorder. Disordered systems are analyzed e.g. in [48]. Any system may be described by a set of the scalar quantities R_j which determine the positions of particles (atoms) in the configuration space. We will deal with one-dimensional systems. The simplest ordered system is described by $R_n = \hbar a$ where $n = 1, 2, 3, \ldots$, and a is a constant. If R_j is completely random we have one-dimensional liquid or one-dimensional glass [48]. Various probability hypotheses on distribution of particles along the chain may be worked out. We will assume a simple case of the random independent values R_j whose distribution along the chain has the same probability density. This corresponds to one-dimensional gas. The system may be described by one statistical parameter; we may use the density of particles or the mean distance between particles given by

$$a = \frac{L}{N} \tag{3.32}$$

where L is the length of the chain and N is the number of particles. We

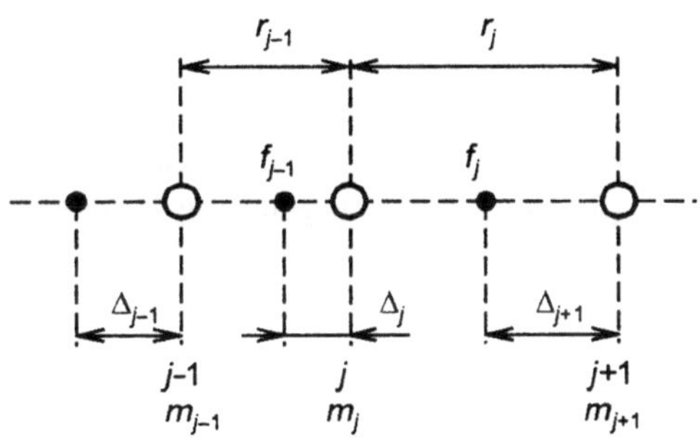

Fig. 3.5 A disordered one-dimensional chain. The particles have different masses and are not equidistant; they are in the equilibrium positions (the displacements q_{j-1}, q_j, and q_{j+1} from the equilibrium positions are not shown). The full dots indicate the equidistant control points. Δ_j denotes the distance of the equilibrium position of the j-th particle from the corresponding control point.

may determine the theoretical equidistant points $R_n = na$ which can be used as the control points in the chain. The equilibrium points determined by the minimum of the configuration energy are displaced from the control points. The absolute coordinate of a particle in the chain has not a basic importance. The relative distance between particles is more useful for the analysis of vibrations. We may introduce a system of relative coordinates (i.e. the distance between two particles)

$$r_j = R_{j+1} - R_j \, . \tag{3.33}$$

The relative coordinate r_j is a random variable too. We have to remark that in molecular chains the positions of particles are not distributed randomly. Their distribution obeys certain laws but our treatment is based on the theory of probability.

We will analyze vibrations in a one-dimensional chain with no translation symmetry (Fig. 3.5). This chain is characterized by inequality of the mass of the particles ($m_j \neq m_i$), of the elastic force constants ($f_j \neq f_i$), and of the relative distances between the particles ($r_j \neq r_i$) for $j \neq i$, $j = 1, 2, 3, \ldots$, and $i = 1, 2, 3, \ldots$. The condition of the short range interaction between the closest neighbours is assumed. The equation of motion for the j-th particle is given in the form

$$m_j \ddot{q}_j = f_{j-1} q_{j-1} + f_j q_{j+1} - (f_{j-1} + f_j) q_j \tag{3.34}$$

where f_j denotes the elastic force constant corresponding to the relative distance r_j and q_j is the displacement from the equilibrium position of the j-th particle. We assume that the vibrations propagate along the chain as a wave, e.g. in the positive direction. The vibration of a particle in the position j is transmitted to the particle in the position $(j+1)$ and further to the successive particles. Each group of particles contained in Eq. (3.34), i.e. the particles in positions $(j-1)$, j, $(j+1)$, forms an elementary cell mediating transfer of vibrations. For the purpose of vibration transfer we may introduce the transmission matrix [48] which generates successive differential equations of the type given by Eq. (3.34). The corresponding transmission matrix is

$$\begin{bmatrix} q_j \\ q_{j+1} \end{bmatrix} = \begin{bmatrix} 0 \quad , & 1 \\ -\frac{1}{f_j} f_{j-1} \quad , & -\frac{1}{f_j} \left[m_j \omega^2 - (f_{j-1} + f_j) \right] \end{bmatrix} \begin{bmatrix} q_{j-1} \\ q_j \end{bmatrix} \tag{3.35}$$

where $-\omega^2 q_j$ was substituted for $\ddot{q}_j$.

We can write Eq. (3.35) in a symbolic form

$$q_{j+1} = T_j q_j \tag{3.36}$$

which describes transfer of the excitation along the chain and corresponds to the relation (3.34).

In our case the transmission matrix T_j is a matrix 2×2 whose components are given by Eq. 3.35. The excitation propagates from the j-th particle to the $(j+1)$-th particle. Therefore, using Eq. (3.36) one may describe propagation in the chain by the relation

$$q_{j+\ell} = T_{j+\ell-1} T_{j+\ell-2} \ldots T_j q_j \tag{3.37}$$

where ℓ is an integer determining the number of elementary cells.

In an ordered system the transmission matrices may be identical and Eq. (3.37) yields

$$q_{j+\ell} = [T]^{\ell} q_j \ . \tag{3.38}$$

In a disordered system the transmission matrix changes from one elementary cell to the following cell along the chain. We may say that the transmission matrix is a random function of j. The elements of the transmission matrix depend on the physical properties of the chain.

We want to determine the dependence of the frequency on the wave constant in the disordered system. First of all we will analyze the one-dimensional ordered system where the relative distance $r_j \neq r_{j+1}$ but $r_j = r_{j+2}$ and $r_{j-1} = r_{j+1}$ for all j in the chain (the short relative distance between particles alternates with the long distance as is shown in Fig. 3.6). We assume solution of Eq. (3.34) in the form

$$\begin{aligned} q_{j-1} &= A_2 \exp\{-\mathrm{i}[\omega t - k(n-1)a - k\Delta_2)]\} \\ q_j &= A_1 \exp\{-\mathrm{i}[\omega t - kna - k\Delta_1)]\} \\ q_{j+1} &= A_2 \exp\{-\mathrm{i}[\omega t - k(n+1)a - k\Delta_2)]\} \end{aligned} \tag{3.39}$$

where j and n corespond one another, A_1, A_2 are the amplitudes, and Δ_2,

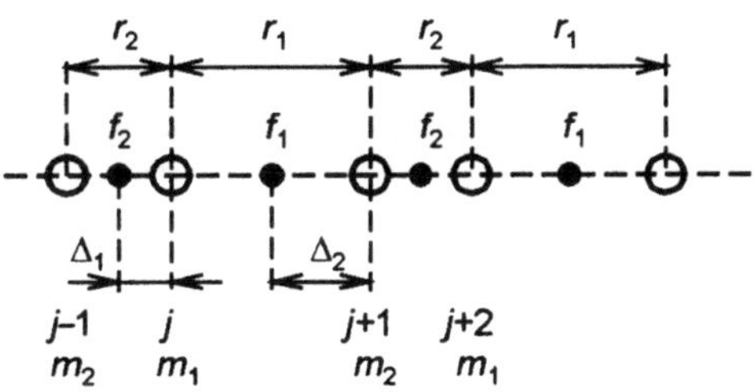

Fig. 3.6 A symmetric one-dimensional chain containing particles with the masses m_1 and m_2. The particles are in the equilibrium positions. The distances between the particles are r_1 and r_2. The full dots indicate the equidistant control points. Δ_1 and Δ_2 denote the distance of the equilibrium points from the corresponding control points.

Δ_1, and Δ_2 are the deviations of the equilibrium positions of the $(j-1)$-th, j-th, and $(j+1)$-th particles from the corresponding equidistant control points $(n-1)a$, na, and $(n+1)a$, respectively (Fig. 3.6). From Eq. (3.34) for $\ddot{q}_j$ and $\ddot{q}_{j+1}$ after substitution q_{j-1}, q_j, q_{j+1}, and q_{j+2} (from Eqs (3.39)) one gets

$$\begin{aligned} &\left[m_1\omega^2 - (f_2 + f_1)\right] A_1 \\ &\qquad + \{f_2 \exp[\mathrm{i}k(-a + \Delta_2 - \Delta_1)] + f_1 \exp[\mathrm{i}k(a + \Delta_2 - \Delta_1)]\} A_2 = 0 \\ &\left[m_2\omega^2 - (f_1 + f_2)\right] A_2 \\ &\qquad + \{f_1 \exp[\mathrm{i}k(-a + \Delta_1 - \Delta_2)] + f_2 \exp[\mathrm{i}k(a + \Delta_1 - \Delta_2)]\} A_1 = 0 \end{aligned} \tag{3.40}$$

where $f_j = f_{j+2} = f_1$, $f_{j-1} = f_{j+1} = f_2$, $m_j = m_{j+2} = m_1$, $m_{j-1} = m_{j+1} = m_2$, $A_j = A_{j+2} = A_1$, and $A_{j-1} + A_{j+1} = A_2$. Solution of Eqs (3.40) exists if the determinant of the matrix of the coefficients at A_1 and A_2 vanishes. The determinant of the matrix equals zero, i.e.

$$\begin{vmatrix} m_1\omega^2 - (f_1 + f_2) & , & \begin{matrix} f_1 \exp[ik(a + \Delta_{21})] \\ +f_2 \exp[ik(-a + \Delta_{21})] \end{matrix} \\ \begin{matrix} f_1 \exp[ik(-a + \Delta_{12})] \\ +f_2 \exp[ik(a + \Delta_{12})] \end{matrix} & , & m_2\omega^2 - (f_1 + f_2) \end{vmatrix} = 0 \qquad (3.41)$$

(where $\Delta_{12} = -\Delta_{21} = \Delta_1 - \Delta_2$). From the determinant given by Eq. (3.41) we obtain

$$\omega^4 - (f_1 + f_2)\frac{m_1 + m_2}{m_1 m_2}\omega^2 + 4\frac{f_1 f_2}{m_1 m_2}\left[1 - \cos^2(ka)\right] = 0\,. \qquad (3.42)$$

If $f_1 = f_2$ then Eq. (3.42) is identical with Eq. (3.27) for an ordered chain with the equidistant positions of particles. The properties of both the types of chains are similar. We can evaluate the dispersion relation which would show two branches of vibrations with ω_+ and ω_-.

We may generalize the procedure given by Eqs (3.39)–(3.42). We adopt a model of N different particles whose positions are not equidistant (Fig. 3.5). The deviation of the j-th particle equilibrium position from the corresponding equidistant control point is Δ_j. We assume that Born and von Karman cyclic conditions are satisfied, i.e. the N-th particle is the neighbour of the first particle. The system of N particles is described by a set of equations which can be written in the form

$$\begin{aligned} &\left[m_1\omega^2 - (f_N + f_1)\right] A_1 + \{f_N \exp[ik(-a + \Delta_{N1})]\} A_N \\ &\qquad + \{f_1 \exp[ik(a + \Delta_{21})]\} A_2 = 0 \\ &\left[m_2\omega^2 - (f_1 + f_2)\right] A_2 + \{f_1 \exp[ik(-a + \Delta_{12})]\} A_1 \\ &\qquad + \{f_2 \exp[ik(a + \Delta_{32})]\} A_3 = 0 \\ &\qquad \vdots \\ &\left[m_j\omega^2 - (f_{j-1} + f_j)\right] A_j + \left\{f_{j-1} \exp\left[ik(-a + \Delta_{(j-1)j})\right]\right\} A_{j-1} \\ &\qquad + \left\{f_j \exp\left[ik(a + \Delta_{(j+1)j})\right]\right\} A_{j+1} = 0 \\ &\qquad \vdots \\ &\left[m_N\omega^2 - (f_{N-1} + f_N)\right] A_N + \left\{f_{N-1} \exp\left[ik(-a + \Delta_{(N-1)N})\right]\right\} A_{N-1} \\ &\qquad + \{f_N \exp[ik(a + \Delta_{1N})]\} A_1 = 0 \end{aligned} \qquad (3.43)$$

where $\Delta_{(j-1)j} = \Delta_{j-1} - \Delta_j$ and similar relations hold for the other terms of this type.

If we denote

$$\alpha_{(j+1)j} = \exp\left[ik(a + \Delta_{j+1} - \Delta_j)\right]$$

and its complex conjugate

$$\alpha^*_{(j+1)j} = \exp\left[ik(-a - \Delta_{j+1} + \Delta_j)\right]$$

then the matrix of the coefficient at the amplitudes $A_1, A_2, \ldots, A_j, \ldots, A_N$ of the set of equations (3.43) may be written as

$$\begin{bmatrix} m_1\omega^2 - (f_N + f_1) & f_1\alpha_{21} & 0 & 0 & \cdots & f_N\alpha^*_{1N} \\ f_1\alpha^*_{21} & m_2\omega^2 - (f_1 + f_2) & f_2\alpha_{32} & 0 & \cdots & 0 \\ 0 & f_2\alpha^*_{32} & m_3\omega^2 - (f_2 + f_3) & f_3\alpha_{43} & \cdots & 0 \\ \vdots & \vdots & \vdots & \vdots & & \vdots \\ f_N\alpha_{1N} & 0 & \cdots & 0 & f_{(N-1)}\alpha^*_{N(N-1)} & m_N\omega^2 - (f_{N-1} + f_N) \end{bmatrix} \tag{3.44}$$

The matrix of the coefficients given by Eq. (3.44) is Hermitian. The roots of the determinant of a Hermitian matrix are real. Therefore, the angular frequencies ω are real. A solution exists if the determinant of the matrix given by Eq. (3.44) is equal to zero. We get N values of ω which correspond to the N branches of frequencies (the N branches of vibrations) in a chain with the translation symmetry. There is a principal difference between the solution given by Eq. (3.44) and the solution for a chain with translation symmetry. In a symmetric chain N may be arbitrarily chosen and the values of the wave constants have quasi-continuous distribution. In a disordered chain the number N is fixed and the frequencies ω are given as constants of the disordered system. Regardless of the difference the concept of the vibration modes may be applied to disordered systems too.

We analyzed the system with the elastic force constants

$$f_1 \neq f_2 \ldots f_{j-1} \neq f_j \neq f_{j+1} \ldots f_{N-1} \neq f_N$$

where the potential of the particles as a function of distance is asymmetric with respect to the equilibrium position (a similar type of asymmetry has the Morse potential). Therefore, the system is non-linear and the spectral energy transfer exists.

3.7 Properties of the Cytoskeleton

Before we will analyze the vibrations we have to describe the main properties of microtubules and other structures in the cytoskeleton. The cytoskeleton is a highly dynamic structure which reorganizes continually as the cell changes its shape, divides, and responds to its environment [1]. The cytoskeleton is composed of intermediate filaments, actin filaments (which are also called microfilaments), and of microtubules. The filaments and the microtubules are mutually connected and form a three dimensional network in the cell. The centre of the cytoskeleton is the centrosome which is assumed to be its major organizing centre. On the periphery the cytoskeleton is connected to the transmembrane proteins. The cytoskeleton organizes the cell, mediates transport of molecules, structures, organelles, synaptic vesicles, and other cell components. The microtubules in the cytoskeleton are used by motor proteins as a "highway" for "cargo" transport. The motor proteins of two classes (the kinesins and the cytoplasmic dyneins) are bound to the microtubules and move along them. The cytoskeleton receives signals from the cellular environment mediated by the membrane proteins and participates in signal transmission outwards from the cell. It has a fundamental role in the mitotic phase forming the mitotic spindle. The mitotic spindle is an array of microtubules "radiating" from the centrosomes positioned on the opposite sides in a dividing cell.

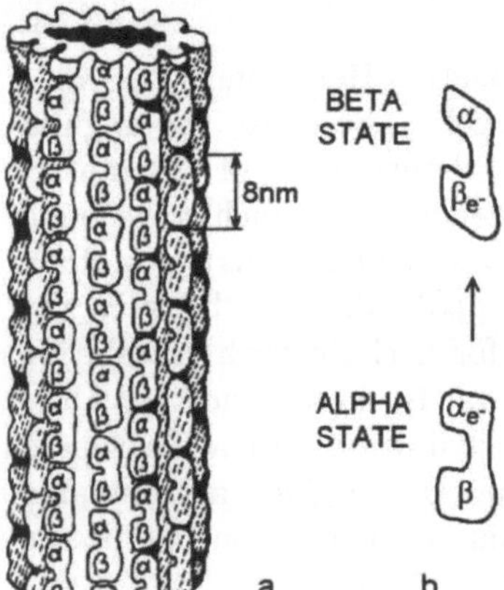

Fig. 3.7 a) A schematic illustration of a microtubule with the tubulin molecules aligned into parallel rows – protofilaments. Each of the thirteen protofilaments forming the microtubule is composed of series of tubulin heterodimers. b) Conformation change of a tubulin heterodimer from the α state to the β state is shown. After Amos and Klug [49] and Tuszyński et al. [50].

The intermediate filaments are ropelike fibres with a diameter of about 10 nm. They are composed of fibrous molecules (vimentins, keratins, nuclear lamins, neurofilament proteins, etc.). Their main role is connected with resistance to mechanical stresses.

The actin filaments are two stranded helical polymers with a diameter of about 8 nm. They are composed of a globular protein actin. The actin filaments are organized in linear bundles, two dimensional networks, and in three dimensional gels. They are dispersed throughout the cell but mainly concentrated in the cell cortex which controls the shape and surface movements of most animal cells. They are polarized and their function depends on polarity.

The microtubules, long polymers extending throughout the cytoplasm are the main organizers of the cytoskeleton; they are highly dynamic structures [1,51–54]. The microtubule resembles a cylindrical tubular structure with a diameter of 25 nm and a central core which is very likely empty (filled with cytosol). The cylindrical structure can be considered as being built from 13 (or 14) protofilaments packed in parallel to form a wall of the cylinder (Fig. 3.7a) [49]. The protofilaments are one-dimensional chains of heterodimer tubulin molecules. The tubulin molecule is composed of two globular peptides – α- and β-tubulins. The dimensions of these monomers are about 4 nm; each of them has the relative molecular mass of about 55 000 ("molecular weight" in daltons) but the masses are not equal. The heterodimer molecule can change conformation from the α state (non-tilted) to the β state (tilted by 29 degrees from the microtubule axis [55]) – Fig. 3.7b. This conformation transition of the heterodimer subunits tends to force the protofilament into a curved shape that is less able to pack into the microtubule wall.

The microtubules are polar with the plus and the minus end [1,56]. Each tubulin heterodimer in the microtubule is a polar subunit whose length is of about 8 nm. The heterodimer may be physically viewed as an electric dipole [50,57,58] whose dipolar character originates from 18 calcium ions bound to each β monomer. An equal number of the negative charges required for electrostatical balance are positioned near the neighbouring α-monomer. A strong electric field is present not only inside the heterodimer but also btween adjacent heterodimers causing the Coulomb attractive interaction. The dipole–dipole energy of a central dipole interacting with six of its neighbours amount to approximately 9.62×10^{-21} J in the non-tilted case and 6.94×10^{-21} J in the tilted case [50].

As a result of the dipole nature of the heterodimers the microtubules are polar structures with the fast growing plus end on the periphery and the slowly growing minus end usually embedded in the centrosome (the majority of the microtubules nucleate in the centrosome). If the charges at the ends are not compensated the microtubule cylinder taken as a whole represents a giant dipole so that the microtubule generates an intrinsic electric field parallel to microtubule's axis. Additional potential energy of the β-tubulin is

$$U_{\text{el}} = -geEq_n \tag{3.45}$$

where ge is the positive charge in the β-tubulin monomer, E is the intensity of the intrinsic electric field, and q_n is the longitudinal displacement of the n-th constituent heterodimer. For the α-tubulin with the negative charge U_{el}

has the opposite sign. Polarity of the microtubules are closely connected with their function [59,60]. The function of the microtubules can be regulated by phosphorylation and dephosphorylation of microtubule-associated tau protein [61].

Guanosine triphosphate molecules (GTP) are bound to both tubulins in the heterodimer. After polymerization when the heterodimer is attached to the microtubule the GTP bound to the β-tubulin is hydrolyzed to the guanosine diphosphate (GDP). The α-tubulin GTP cannot be hydrolyzed and is a fixed part of the tubulin structure. The rate of polymerization may be greater than the rate of hydrolysis and an energy rich cap (a cap of heterodimers where GTP is not hydrolyzed) can be formed at the end of the tubulin. This cap has an important function. To continue growing at least one layer of the heterodimer molecules at the end of the microtubule should remain non hydrolyzed (i.e. 13 or 14 tubulin molecules) [62]. The energy rich cap at the end of the microtubule is hydrolyzed later. The significance of GTP in microtubule dynamics is analyzed in [52,53,62].

The dynamics of the microtubules belong to their most prominent features. The microtubules are capable of rapid polymerization and depolymerization [51] which enable exchange of the subunits between soluble and polymer pools. The microtubules grow towards the cell periphery and then shrink back towards the centrosome. They may shrink partially and then grow again or they may disappear completely to be replaced by new microtubules. This process is called dynamic instability. The treadmilling (i.e. addition of the tubulin subunits at the plus end and removal of the tubulin subunits at the minus end at an identical rate) appears if both the ends of the microtubule are exposed [1]. The dynamics of the microtubule depend on the energy rich cap at its end and may depend on kinetochores too [63].

The microtubules depolymerize and repolymerize continually in animal cells. The majority of the microtubules can incorporate new subunits within 20–30 min although complete exchange of the subunits in a cell can be achieved within 1 hour [54]. The half-life of an individual microtubule is of about 10 min [1]. The depolymerization removes the subunits with GDP and repolymerization adds the energy rich subunits with GTP to a microtubule. The hydrolysis energy is not used for polymerization. The majority of the hydrolysis energy is stored in the microtubule [64]. The mechanism of the dynamic instability (depolymerization and repolymerization) provides continual supply of energy into the microtubule structure in a cell and signifies that the total energy storage is renewed within a time period shorter than one hour as this mechanism can add and remove the heterodimers without any effect on the total pool of the heterodimers in the microtubules.

The dynamic instability of the microtubules has another aspect too. Tubulin subunits can be covalently modified after they polymerize. These modifications on the α-tubulin are connected with acetylation on a particular lysine and the removal of the tyrosine residue from the carboxyl terminus of the α-tubulin. The level of these modifications increases with time forming a molecular clock inside the tubulin structure [1].

The mechanism of energy supply by hydrolysis of an energy rich compound is also functioning in the actin molecules added to the actin filament (in this case adenosine triphosphate – ATP is hydrolysed to adenosine diphosphate – ADP). This mechanism of hydrolysis is similar to that in the microtubules.

Conformation changes of tubulin heterodimers may be described in terms of the double well potential model (Fig. 3.8). The minimum values of the potential of both wells may be the same (Fig. 3.8a) or the minimum values may be on different levels with the potential difference ΔU (Fig. 3.8b). The potential in the former case may be described by the relation [65,66]

$$U_n = -\frac{A}{2}u_n^2 + \frac{B}{4}u_n^4 \tag{3.46}$$

(with a potential barrier $|A^2/2B|$ between two valleys) and the potential in the latter case may be given by

$$U_n = -\frac{A}{2}u_n^2 + \frac{B}{4}u_n^4 + \frac{C}{3}u_n^3 \tag{3.47}$$

where u_n is the conformation coordinate and A, B, C are the constants.

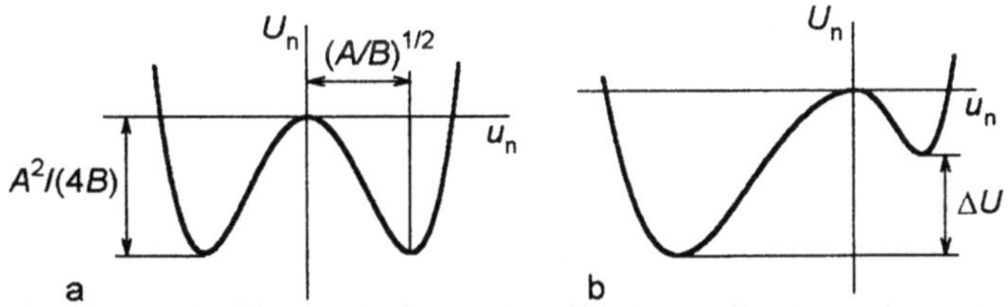

Fig. 3.8 A schematic illustration of the double well potential models for two states of tubulin heterodimer conformation.

The two potential valleys can have different elastic force constants as is shown in Fig. 3.8b. The potential valleys need not be symmetric with respect to the potential minima. Energy supply from hydrolysis of GTP can switch over the conformation from the α state to the β state. If the minimum values of the potential in both valleys are the same (Fig. 3.8) the whole energy used for surmounting the potential barrier between valleys may be transferred to the vibration states. If the minimum values of the potential are on different levels (e.g. the β state is on a higher energy level than the α state – Fig. 3.8b) a part of the energy is expended on the energy difference ΔU and only the remaining part of the energy supplied may be converted into energy of vibrations after the transition from the α to the β state.

The cytoskeleton is an organizing system of the cell and exerts forces and generates movements without any major chemical changes [1]. The mechanisms of the cytoskeleton activity and the way of energy utilization are still not well understood. Physical mechanisms are probably fundamental to the cytoskeleton function. It seems reasonable to assume that the cytoskeleton can use the energy from hydrolysis of ATP and GTP stored in the actin filaments and in the microtubules, respectively. We will analyze vibrations in the microtubules and in the actin filaments as they may be one of possible mechanisms underlying the cytoskeleton function.

3.8 Thermal Vibrations in Microtubules and Actin Filaments

The heterodimers in a protofilament form a one-dimensional chain. The globular tubulins in the chain may be considered as mass units with translation symmetry. These mass units have internal vibrations (exerted by atoms and by large parts of tubulins) as well as external vibrations in the chain (i.e. vibrations exerted by each tubulin as a whole). A similar rationale may be applied to the actin filaments. We will analyze the external vibrations in the protofilaments and in the actin filaments using a method similar to that employed in the solid state physics. Nevertheless, we have to point out certain differences between crystalline solids with translation symmetry and the microtubules (or the actin filaments). The "diameters" of atoms in solids are of about 0.1 nm and the distances between their centres are about twice or three times greater. The "diameters" of the tubulins and of the actins are of about 4 nm; they are only slightly smaller than the distances between their centres of mass. In contrast with atoms the tubulin and actin globules are not rigid but rather elastic particles. Their elastic deformations can change the external vibrations and mediate interaction between them and the internal vibrations. First of all we will analyze the external vibrations without including dynamic deformations of the tubulin and of the actin globules.

A scheme of the one-dimensional chain is shown in Fig. 3.6. In this case the empty circles denote the centres of mass of the tubulins in equilibrium positions. The vibration entities are the globular tubulin monomers, i.e. the α- and the β-tubulins with masses m_1 and m_2, respectively. The elastic force constants between the α- and the β-monomers inside the tubulin heterodimer and between two neighbouring heterodimers are f_1 and f_2, respectively.

Analysis of the vibrations is similar to that for the random vibrations in the crystalline structures with translation symmetry. We will derive the dispersion relation for a protofilament. The problem is identical with that described by Eqs (3.39)–(3.42). From Eq. (3.42) we obtain

$$\omega_{\pm}^2 = \frac{(f_1+f_2)(m_1+m_2)}{2m_1m_2} \pm \sqrt{\left[\frac{(f_1+f_2)(m_1+m_2)}{2m_1m_2}\right]^2 - 4\frac{f_1f_2}{m_1m_2}\sin^2(ka)}\,. \tag{3.48}$$

A similar result is given by Eq. (3.28) for an ordered chain with particles of two types. The optical (ω_+) and acoustical (ω_-) branches of vibrations exist in the chain. As the monomer masses (m_1 and m_2) are several orders greater than the mass of a single atom and the elastic force constants (f_1 and f_2) are very likely considerably smaller than similar quantities in a crystalline structure the frequencies of vibration modes are low. The bond between the α- and the β-monomers inside a heterodimer is very likely stronger than the bond between the neighbouring heterodimers and, therefore, $f_1 > f_2$.

We will analyze vibrations in a microtubule using Eq. (3.48). The protofilaments in a microtubule are bound and build up a wall of a cylinder. We

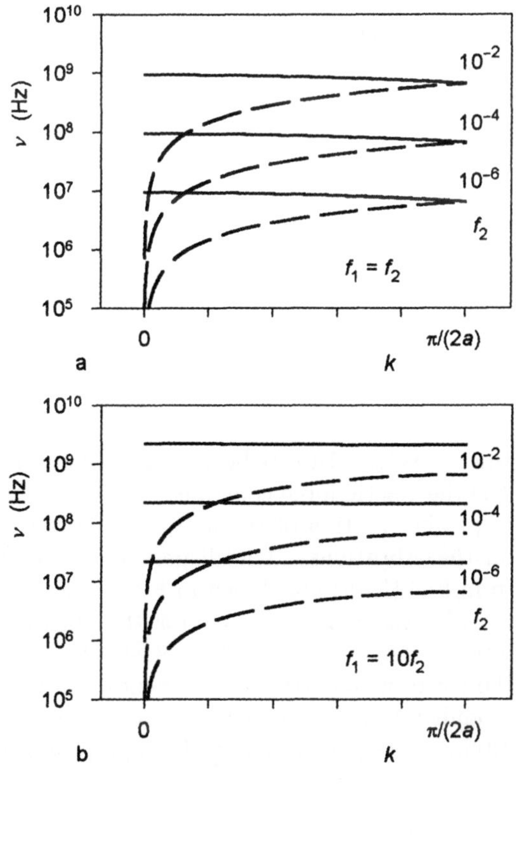

Fig. 3.9 Dispersion relation for the microtubule (the frequency of vibrations as a function of the absolute value of the wave vector k). The full curves – the optical branches, the dashed curves – the acoustical branches. The elastic force constants are assessed:
a) $f_1 = f_2$, b) $f_1 = 10f_2$.

will assume that the vibrations in particular protofilaments in the microtubule are not independent and that they are mutually coupled forming a common system of vibrations. As the microtubule is composed of 13 protofilaments we assume, that a microtubule "particle" has 13 times greater mass than a protofilament "particle". (This may be acceptable only as a zero order approximation; the protofilaments in the microtubule are shifted so that the tubulin heterodimer positions do not coincide.) The elastic force constants in isolated parts of the protofilaments (oligomers) and in the microtubules may be different. The frequency versus the absolute value of the wave vector evaluated from Eq. (3.48) is shown in Fig. 3.9a for $f_1 = f_2$ and in Fig. 3.9b for $f_1 = 10 f_2$. The elastic force constants are parameters of the curves.

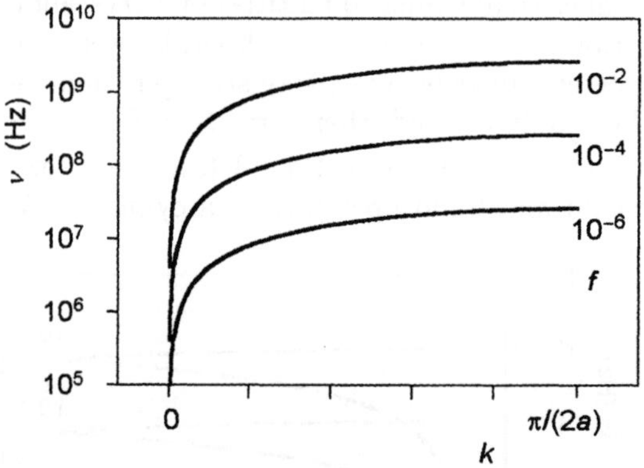

Fig. 3.10 Dispersion relation for the actin filament.

The actin filaments are composed of identical molecules – actins – in a double stranded organization. We may, therefore, apply a simple model of translation symmetry with the equidistant particles. The dispersion relation is given by Eq. (3.8)

$$\omega = 2\sqrt{\frac{f}{m}}\left|\sin\left(\frac{ka}{2}\right)\right| \tag{3.49}$$

where a is the distance betwen the centres of the particles, m is their mass, and f is the elastic force constant. The external vibrations have only one branch. Reptating motion of the actin filaments [67] might depend (to some degree) on the vibrations. The dispersion curves for the actin filaments evaluated from Eq. (3.49) are shown in Fig. 3.10.

We used several values for the elastic force constants to show their effect on the frequency. The elastic force constants might be assessed from the shear strain-stress experimental data [68–70]. The dynamic elastic shear moduli for microtubules and for actin filaments are of about $5\,\mathrm{Nm}^{-2}$ and $20\,\mathrm{Nm}^{-2}$, respectively [69]. Using relation between the elastic shear modulus and Young's modulus and assuming the distance between the neighbouring particles of about 4 nm the elastic force constants might be in the range of 10^{-5}–10^{-6} Nm^{-1}. Nevertheless, filaments in the cytoskeleton have different stiffness under different conditions. The large elastic modulus appears to depend on the fact that filaments are arranged orthogonally and cross-linked to each other at junction points. Changes of viscoelasticity very likely come about by changing the nature of the cross-links or the filament length. The elastic moduli for actin filaments in entangled ensembles are assessed as high as 10^4 Nm^{-2} [71,72]. The corresponding elastic force constants may be greater than 10^{-4} Nm^{-1}.

For a microtubule the lowest frequency for $\nu \neq 0$ of the acoustical branch is in the band of 5–50 MHz if the length of a microtubule is 10 µm ($= \lambda/2$) and velocity of the acoustical waves is in the range from 10^2–10^3 ms^{-1}. The mass of the "particles" in the actin filaments is smaller than in the microtubules and, therefore, the frequencies are greater. The lowest frequencies of vibrations of the acoustical branches in the microtubules as well as in the actin filaments are in the classical radio bands from 1 MHz to 1 GHz which corresponds to the value determined by Pohl [27] from experimental findings based on dielectrophoretic effects of cells. The frequency of yeast cell emission was assessed to be of the order of 1 MHz.

The values derived by the present calculations presumably will be modified in biological reality because a number of simplifications was used, for example the internal vibrations were neglected and simple structures of the microtubules and of the actin filaments were assumed. The tubulins as well as the actins are rather elastic than rigid globules which may have considerable influence on their vibrations. The internal vibrations are combined with the external vibrations giving rise to complicated spectral patterns. On account of small mass and great values of the elastic force constants the frequencies of the internal vibrations are much higher than those of the external vibrations. The frequency spectrum can contain frequency lines in the millimeter and in the far infrared region too.

If the elastic force constants in a tubulin heterodimer and between two tubulin heterodimers are not equal (i.e. if $f_1 \neq f_2$) the potential of "the tubulin particle" as a function of distance is asymmetric with respect to the equilibrium position The system is non-linear and the higher harmonic components with the frequencies $\omega = n\omega_j$ and the components with the combination frequencies $\omega = n\omega_j + m\omega_k + \ldots$ are generated (n, m are integers). The dispersion relation (3.48) does not contain these harmonic and combination terms.

The heterodimers in the neighbouring microtubule protofilaments are mutually displaced along the microtubule axis which can cause phase shifts and alter the types of vibrations: torsion and/or helicoid vibrations can exist in the microtubule. The heterodimers in the microtubule are organized in hexagons spanning the helical spiral on the circumference of the microtubule. Some proteins from the family of microtubule associated proteins (MAP's) interconnect the microtubules, the intermediate filaments, and the actin filaments and create a common vibration sytem of the cytoskeleton.

3.9 Energy Supply to the Cytoskeleton

When the microtubule grows tubulin heterodimers add to the free plus end and the GTP molecules attached to the β-tubulins are hydrolyzed to the GDP's [1,52,53,62,64] except for the energy rich cap at the end of the microtubule which is hydrolyzed later. The measurements of hydrolysis of the GTP analogue guanylyl–(a,b)–methylene–diphosphonate (GMPCPP) in solution, in tubulin heterodimers, and in tubulin subunits in the microtubule led to a conclusion that the free energy was −5.18, −3.79, and −0.9 kcal/mole (−21.7, −15.9, and −3.8 kJ/mole), i.e. −0.22, −0.16, and −0.04 eV per one entity, respectively [64]. Therefore, energy of about 4 kcal/mole (17 kJ/mole) was assumed to be potentially available from the hydrolysis of GTP. A part of the energy is expended on dissociation of the inorganic phosphate P_i but the majority of it is stored in the microtubule lattice [64]. A remeasurement of the dissociation rate constant allowed better estimation of the free energy for hydrolysis and release of P_i which is +0.7 kcal/mole (+2.9 kJ/mole) [73] rather than −0.9 kcal/mole as was reported in [64]. The energy stored in the microtubule lattice is assessed to be 1.7 kcal/mole (7.1 kJ/mole). We have to emphasize that the measurements of the energy available from hydrolysis were carried out on the GTP analogue GMPCPP.

Similar mechanisms may be found in the actin filaments. The actin molecule is clam shaped with the ATP molecules bound in the crevice. ATP in the actin molecule added to the filament is hydrolyzed. Hydrolyzation and formation of an energy rich cap are analogous to the processes in the microtubules. Hydrolysis of ATP can release and supply energy up to about 13 kcal/mole (54 kJ/mole).

The experimental findings given e.g. in [64,73] show that the energy from hydrolysis is stored in the microtubules. The form in which the energy is stored in the microtubule lattice is a special issue. Mandelkow et al. [53] assume that the tubulin–GDP subunit prefers a conformation which leads to coiling of the protofilaments and as a consequence energy may be stored in the form of a tense state of the microtubule lattice. Caplow et al. [64] come to the conclusion that energy is stored in protein conformation and presumably as a repulsive force between the subunits. A similar mechanism of energy storage in the microtubules is given in [50], i.e. that energy from the hydrolysis of GTP to GDP is imparted into the microtubule lattice via tubulin conformation changes. Real physical mechanisms of energy transformation are irreversible and a part of the energy is always transferred to the random heat vibrations. Even if the energy levels of two conformation states of a tubulin subunit are different (e.g. if the level of the β state is higher than that of the α state) after the transition to the higher energy level a part of the energy which is used to surmount the potential barrier between the conformation states can be transformed into the vibration energy. The opposite relaxation process and restoration of the α conformation state should convert energy to the microtubule vibrations too. Creation of the tense state with altered forces between the subunits in the microtubule has to supply energy into the lattice. Therefore, we may assume that energy is stored in the microtubule lattice not only in the conformation changes and in the tense state of the lattice but also in the vibration states. Besides the energy stored in the cytoskeleton during polymerization a part of the energy from the hydrolysis of ATP supplied to motor proteins moving along a microtubule [1] can be transferred to the microtubule as the random excitations of the vibrations.

TABLE 1 Relative value of the energy E/E_0 per one entity supplied from GTP and ATP

Frequency (Hz)	E/E_0	
	GTP (4 kcal/mole)	ATP (10 kcal/mole)
10^7	6.7	16.8
10^{11}	6.8	16.9

Energy stored in the microtubule can do work in the cell [64] but a precise manner in which the energy is used is still not understood [74]. As the cytoskeleton is not subject to any major chemical changes [1] the energy has to be used in some biophysical processes. The hydrolysis energy can change bond strength [74] and excite solitons or soliton like waves [57,74]. It causes the conformation changes of the tubulin heterodimers and, therefore, kink wave reorientation of the heterodimer dipoles can be excited in the microtubule [57]. It can excite the lattice vibrations and polarization waves, which may be coherent on account of the Fröhlich mechanism [74] or of the intrinsic non-linear mechanism. (The Fröhlich mechanism is described e.g. in [7,8,75]). Regardless of the primary mechanisms of the energy storage at a later stage at least a part of it is always supplied to the vibrations in the microtubules. Nevertheless, the majority of the energy stored in the microtubules may be converted into the energy of vibrations. Therefore, the vibrations are excited and can be far from thermodynamic equilibrium. We suggest that the mechanism of the coherent polar vibrations in the microtubules using the stored energy is a tool for the microtubule organization function. The electric field generated by the polar vibrations can do work in the surroundings of the microtubules. A mechanism of this type (large scale quantum coherent phenomena in microtubules) is suggested and supported by theoretical analysis e.g. in [76,77]. The mechanism discussed e.g. in [76] assumes interaction between coherent modes in the microtubule with water in its surroundings, especially in the microtubule cavity. Experimental data necessary for assessment of the energy transfer between the microtubule and its ambient medium and of the damping of the vibrations are still missing.

We assess the amount of energy which might be supplied to the cytoskeleton vibrations from the hydrolysis of GTP and ATP. We evaluate relative values of the energy E/E_0 per one entity where E is the energy supplied from the hydrolysis of GTP or ATP and E_0 is the energy of an oscillator in thermodynamic equilibrium in the frequency range from 10 MHz to 100 GHz, which include the possible range of external vibrations. If the total energy supplied from the hydrolysis is 4 kcal/mole (10 kcal/mole) then the energy per one entity is about 7 times (17 times) greater than the energy of an oscillator in thermodynamic equilibrium (Tab. 1). Due to the non-linear mechanism the energy may be accumulated in certain vibration modes. We note that on account of the dynamic instability the total energy supplied to the microtubule structure in a cell within one hour can be much greater than the energy corresponding to the number the heterodimers in the microtubule structure. This energy can play a basic role in the cytoskeleton activity.

Before we will end the discussion about the energy supply to the cytoskeleton we emphasize that the thermal vibrations are an omnipresent phenomenon and living matter is not any exception. We have also to note that all real processes in nature are irreversible: a part of the energy used in the processes is lost and transformed into vibrations. In non-linear systems the final thermalization of this part of energy can go through the excitation of certain vibration modes, i.e. through the excitation of a state which is above thermodynamic equilibrium. Therefore, production of entropy in non-linear systems has very likely certain special features.

3.10 Hamiltonian of Vibrating Particles in One-Dimensional Chain

We will examine a model one-dimensional system similar to that in Fig. 3.1 [78]. We will write the Hamiltonian of the system containing discrete equidistant particles with a unit mass. For the sake of simplicity we assume that the distance between the neighbouring particles $a = 1$, the elastic force constant $f = 1$, and Planck's constant $\hbar = h/2\pi$ is normalized to 1 too. The displacements of particles from their equilibrium positions are denoted here by $q_r(t)$. The momentum coordinate is $p_r(t)$. The Lagrangian is given by

$$L = T - U = \frac{1}{2}\sum_r \dot{q}_r^2 - \frac{1}{2}\sum_r (q_{r+1} - q_r)^2 \tag{3.50}$$

where T is the kinetic energy, U is the potential energy and

$$p_r = \frac{\partial L}{\partial \dot{q}_r} = \dot{q}_r$$

(where $\dot{q}_r$ is the first derivative of q_r with respect to time). The Hamiltonian function is given by

$$\begin{aligned} H &= \sum_r p_r \dot{q}_r - L = \sum_r p_r \dot{q}_r - \frac{1}{2}\sum_r (\dot{q}_r)^2 + \frac{1}{2}\sum_r (q_{r+1} - q_r)^2 \\ &= \frac{1}{2}\sum_r (\dot{q}_r)^2 + \frac{1}{2}\sum_r (q_{r+1} - q_r)^2 \ . \end{aligned} \tag{3.51}$$

An operator is associated with every physical observable (e.g. in our case with q_r, p_r, L, H) which makes possible to determine eigenvalues of the observable. We will not use special symbols for the operators; the symbols q_r, p_r, Q_r, P_r, L, and H will denote both physical observables as well as operators. We assume validity of the commutation relations for the operators q_r and p_s in the form

$$\begin{aligned} [q_r, p_s] &= q_r p_s - p_s q_r = i\delta_{rs} \\ [q_r, q_s] &= q_r q_s - q_s q_r = 0 \\ [p_r, p_s] &= p_r p_s - p_s p_r = 0 \ . \end{aligned} \tag{3.52}$$

We will find the eigenvalues (i.e. the frequencies) of the operator H – given by Eq. (3.51). The Fourier transform of q_r may be written in the form

$$q_r = \frac{1}{\sqrt{N}}\sum_k Q_k \exp(ikr) \ , \quad Q_k = \frac{1}{\sqrt{N}}\sum_s q_s \exp(-iks) \tag{3.53}$$

(where N is the number of particles in the chain and Q_k is the wave coordinate) which enables transition to phonons. We will use the cyclic boundary conditions $q_{r+N} = q_r$. Then $\exp(ikN) = 1$ and $kN = 2\pi g$ for $g = 0, \pm 1, \pm 2, \ldots, \pm\frac{1}{2}N - 1, \frac{1}{2}N$. The wave constant (generally the wave vector) is given by the relation $k = 2\pi g/N$. Using the sum of the geometric series one can easily prove that

$$\sum_r \exp\left[\mathrm{i}(k+k')r\right] = \begin{matrix} \nearrow 0 & \text{for} & k+k' \neq 0 \\ \searrow N & \text{for} & k+k' = 0\,. \end{matrix} \tag{3.54}$$

We will evaluate the Lagrangian and the Hamiltonian function using the Fourier transform given by Eq. (3.53). The first term on the right hand side of Eq. (3.50) is given by

$$\sum_r (\dot{q}_r)^2 = \frac{1}{N}\sum_k \sum_{k'} \sum_r \dot{Q}_k \dot{Q}_{k'} \exp\left[\mathrm{i}(k+k')r\right] = \sum_k \dot{Q}_k \dot{Q}_{-k} \tag{3.55}$$

where Eq. (3.54) was used.

In a similar way we evaluate the second term on the right hand side of Eq. (3.50)

$$\begin{aligned} &\sum_r (q_{r+1} - q_r)^2 \\ &\quad = \frac{1}{N}\sum_k \sum_{k'} \sum_r Q_k Q_{k'} \left[\exp(\mathrm{i}k) - 1\right]\left[\exp(\mathrm{i}k') - 1\right] \exp\left[i(k+k')r\right] \\ &\quad = 2\sum_k (1 - \cos k)\, Q_k Q_{-k}\,. \end{aligned} \tag{3.56}$$

After substitution of Eqs (3.55) and (3.56) into Eq. (3.50) we get

$$L = \frac{1}{2}\sum_k \dot{Q}_k \dot{Q}_{-k} - \sum_k (1 - \cos k) Q_k Q_{-k}\,. \tag{3.57}$$

The operators of physical observables are Hermitian, i.e $Q_k = Q^+_{-k}$. The momentum is defined as

$$P_k = \frac{\partial L}{\partial \dot{Q}_k} = \dot{Q}_{-k} = P^+_{-k}\,. \tag{3.58}$$

Substituting Eqs (3.55), (3.56), and (3.58) into Eq. (3.51) we get

$$H = \frac{1}{2}\sum_k P_k P_{-k} + \sum_k (1 - \cos k) Q_k Q_{-k}\,. \tag{3.59}$$

We introduced new coordinates Q_k, P_k, where

$$\begin{aligned} P_k &= \frac{1}{\sqrt{N}}\sum_s \dot{q}_s \exp(\mathrm{i}ks) = \frac{1}{\sqrt{N}}\sum_s p_s \exp(\mathrm{i}ks) \\ p_r &= \frac{1}{\sqrt{N}}\sum_r P_k \exp(-\mathrm{i}kr)\,. \end{aligned} \tag{3.60}$$

The commutation relation for Q_k, P_k are given by

$$
\begin{aligned}
[Q_k, P_k] &= \frac{1}{N}\left[\sum_r q_r \exp(-\mathrm{i}kr), \sum_s p_s \exp(\mathrm{i}k's)\right] \\
&= \frac{1}{N}\sum_{r,s} [q_r, p_s] \exp[-\mathrm{i}(kr - k's)] \\
&= \frac{1}{N}\sum_r \mathrm{i} \exp[-\mathrm{i}(k-k')r] = \mathrm{i}\delta_{kk'}
\end{aligned} \tag{3.61}
$$

where $[q_r, p_s] = \mathrm{i}\delta_{rs}$.

We will use the transformation

$$
\begin{aligned}
a_k^+ &= \frac{1}{\sqrt{2\omega_k}}(\omega_k Q_{-k} - \mathrm{i}P_k), & a_k &= \frac{1}{\sqrt{2\omega_k}}(\omega_k Q_k + \mathrm{i}P_{-k}) \\
a_{-k} &= \frac{1}{\sqrt{2\omega_k}}(\omega_k Q_{-k} + \mathrm{i}P_k), & a_{-k}^+ &= \frac{1}{\sqrt{2\omega_k}}(\omega_k Q_k - \mathrm{i}P_{-k})
\end{aligned} \tag{3.62}
$$

where $\omega_k = [2(1-\cos k]^{\frac{1}{2}}$ and a_k^+, a_k, a_{-k}^+, a_{-k} are the operators which enable transition to normal modes. The operators a_k^+, and a_k are called the creation and the annihilation operator, respectively. The inverse transformation is given by

$$
Q_k = \frac{1}{\sqrt{2\omega_k}}\left(a_k + a_{-k}^+\right), \quad P_k = \mathrm{i}\sqrt{\frac{\omega_k}{2}}\left(a_k^+ - a_{-k}\right). \tag{3.63}
$$

Using Eq. (3.62) and (3.63) the commutation relation for a_k^+, a_k may be found

$$
\left[a_k, a_{k'}^+\right] = \frac{1}{2\omega_k}\left\{-\mathrm{i}\omega_k [Q_k, P_{k'}] + \mathrm{i}\omega_{k'} [P_{-k}, Q_{-k'}]\right\} = \delta_{kk'}. \tag{3.64}
$$

We will use the inverse transformation given by Eq. (3.63) to transform Eq. (3.59). We will evaluate the terms

$$
\begin{aligned}
&\frac{1}{2}(P_k P_{-k} + P_{-k}P_k) + (1-\cos k)(Q_k Q_{-k} + Q_{-k}Q_k) \\
&\qquad = \omega_k\left[\left(a_k^+ a_k + \frac{1}{2}\right) + \left(a_{-k}^+ a_{-k} + \frac{1}{2}\right)\right].
\end{aligned} \tag{3.65}
$$

Therefore, the Hamiltonian in Eq. (3.59) is transformed into the form

$$
H = \sum_k \omega_k \left(a_k^+ a_k + \frac{1}{2}\right) \tag{3.66}
$$

which is identical to the Hamiltonian for a harmonic oscillator. The operator $a_k^+ a_k = \hat{n}_k$ is the operator of the number of Bose particles. If n_k is the eigenvalue of the operator $\hat{n}_k$ the Schrödinger equation may be written in the form

$$H\Psi = E\Psi = \left(\sum_k n_k \omega_k\right)\Psi \tag{3.67}$$

where Ψ is the eigenfunction of the operator $\hat{n}_k$. n_k is the number of particles (occupation number) in the normal mode with angular frequency ω_k. If n_k is increased by 1 then one phonon with energy ω_k is created (we assumed that $\hbar = 1$). We derived a quantum system of eigenvalue vibrations which is identical with the system of clasical normal modes.

We introduced the formalism of the second quantization for boson fields. A similar formalism is valid for fermions too. It is often stated that the method of second quantisation is not suitable for solving a wide range of problems. But it can give reasonably simple and translucent rules for describing some physical phenomena. The Hamiltonian is very often so far simplified that it can be easily diagonalised and derivation of any term of the perturbation theory series is not difficult either.

4. Quantum Mechanical Theory of Rate Equations

4.1 The Fröhlich Rate Equation

From the physical point of view the Fröhlich system consists of three fundamental parts: a vibration system, a source of energy, and a heat bath (Fig. 4.1) [7,8,75,79–81]. Before we start with developing the microscopic theory we have to describe these three parts.

The vibration system contains a large number of oscillating units. The oscillating units carry charges with the dipole and/or multipole distributions and generate an electromagnetic field which mediates interactions. Random vibrations of atoms, molecules, and various structures are omnipresent. Energy supply may excite vibrations in biological macromolecules (e.g. in proteins) and also in much more complicated polymeric structures containing a large number of molecules (for instance in the cytoskeleton, in the membranes, etc.); excited vibrations in these oscillating units are far from thermodynamic equilibrium [6]. The level of excitation depends on the energy supply from the source of energy coupled to the oscillating units. The energy is transferred from the source to the oscillating units and transformed into the energy of vibrations. The source of energy and the oscillating units need not be separated in space; they may occupy the same region. One part of a molecule may accept the energy released from ATP, transform it and transfer into the vibrations, and simultaneously participate in the vibrations as a part of the oscillating unit. A mitochondrion (or its part) producing energy by synthesizing ATP may oscillate as well. Any part of the surroundings which supplies energy to the oscillating unit represents a source of energy. Nevertheless, the main part of the medium around the oscillating units is in thermodynamic equilibrium. This part of the ambient medium forms the heat bath. On account of coupling to the ambient medium the oscillating units lose energy. The ambient medium in a cell or outside a cell may consist of individual molecules and particles, of large structures containing great amount of molecules, and of liquid medium (of cytosol and intercellular liquid) with electrons, ions, and charged particles. Therefore, the "biological" heat bath is a very complicated system. The real mechanism of energy transfer may include collisions, Coulomb interactions, etc. First of all we will assume that the heat bath contains systems of neutral structures with vibration modes. Therefore, the heat bath can be described by Bose operators. We will use this approximation in the treatment using the perturbation theory. Nevertheless, the heat bath cannot be restricted only to Bose particles. This evidently does not describe the real conditions in the biological systems. If the vibrations are polar their interactions with electrons, ions and charged particles in the surroundings can be important. Therefore, the particles in the heat bath as well as in the source of energy can be fermions too. In the theoretical treatment both the Bose and Fermi particles have to be considered to describe the real conditions in biological systems.

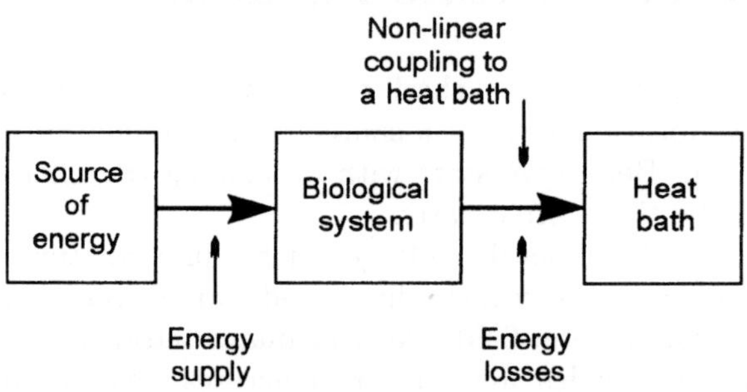

Fig. 4.1 A schematic picture of a Fröhlich system with linear and non-linear coupling to the heat bath.

Coupling of the vibration system to the heat bath is fundamental to the Fröhlich theory. Besides the ordinary one quantum energy transfer between the vibration system and the heat bath Fröhlich assumed also two quanta processes which was originally discussed in connection with electron–phonon interaction in ionic crystals [82]. Analyzing interaction between an electron and continuous dielectric medium with longitudinal polarization waves Fröhlich remarked:

Absorption of a quantum $\hbar\omega$ leads the electron into a level from which after a very short interval it would re-emit a quantum $\hbar\omega$ – in general with different momentum. This suggests to treat such collisions as second order transitions, leading to the absorption of a quantum $\hbar\omega$, followed by a competitive emission of another quantum $\hbar\omega$ whose wave number **w** *can have any value in a whole range.*

Fröhlich was aware of the fact that emission by the electron can excite a vibration mode whose frequency differs from the frequency corresponding to the quantum of energy absorbed by the electron. These two quanta interactions (absorption–emission) lead to extraordinary behaviour of the system.

Fröhlich postulated a rate equation describing the system with two quanta interactions. Nonetheless, connection of the macroscopic quantities in the rate equation with the physical microprocesses remained an open question.

The first attempt to develop a microscopic theory of the Fröhlich coherent vibrations was carried out by Bhaumik et al. in [83,84]. The main contributions were worked out by Wu and Austin [85–88] and Mills [89–95]. Various aspects of the Fröhlich theory were analyzed, e.g. collective behaviour of the biological systems [96,97], self-focusing of the Fröhlich waves [98], and connection of the Fröhlich coherent vibrations and the Davydov solitons [99,100].

The interaction between the oscillating units mediated e.g. by the electromagnetic field produces oscillations in a band of frequencies. A similar rationale may be applied to normal modes. We assign the creation operator a_i^+ and the annihilation (destruction) operator a_i to each mode. The corresponding angular frequency of the mode is ω_i. The angular frequencies of the vibration modes in the source of energy and in the heat bath are Θ_g and Ω_k, respectively. The creation operator P_g^+ and the annihilation operator P_g are assigned to each normal mode in the source of energy which supplies energy to the vibration system. Each heat bath mode is associated with the creation operator b_k^+ and with the annihilation operator b_k. The Hamiltonian of the Fröhlich system may be written in the form [87]

$$H = H_0 + H_1 \tag{4.1}$$

$$H_0 = \sum_i \hbar\omega_i a_i^+ a_i + \sum_k \hbar\Omega_k b_k^+ b_k + \sum_g \hbar\Theta_g P_g^+ P_g \tag{4.2}$$

$$\begin{aligned} H_1 = & \sum_i \sum_k \left(\varphi a_i b_k^+ + \varphi^* a_i^+ b_k\right) + \sum_g \sum_i \left(\zeta P_g a_i^+ + \zeta^* P_g^+ a_i\right) \\ & + \sum_j \sum_i \sum_k \left(\kappa a_j^+ a_i b_k^+ + \kappa^* a_j a_i^+ b_k\right) \end{aligned} \tag{4.3}$$

where φ, ζ, and κ are the coupling constants for the energy transfer between the vibration modes of the vibration system and the heat bath (one quantum energy transfer), for the energy transfer from the source of energy to the vibration modes (one quantum energy supply), and for the energy transfer between two vibration modes of the vibration system with participation of the heat bath (two quanta process), respectively. The Hamiltonian is called the Wu–Austin Hamiltonian and we use it here in the original "historical" form. We assume that all the operators (i.e. a_i^+, a_i, b_k^+, b_k, P_g^+, P_g) operate on bosons. The commutation relations are the following

$$\begin{gathered} \left[a_i, a_{i'}^+\right] = a_i a_{i'}^+ - a_{i'}^+ a_i = \delta_{ii'}, \quad \left[b_k, b_{k'}^+\right] = b_k b_{k'}^+ - b_{k'}^+ b_k = \delta_{kk'} \\ \left[P_g, P_{g'}^+\right] = P_g P_{g'}^+ - P_{g'}^+ P_g = \delta_{gg'} \\ \left[a_i, a_{i'}\right] = 0, \quad \left[b_k, b_{k'}\right] = 0, \quad \left[P_g, P_{g'}\right] = 0 \\ \left[a_i^+, a_{i'}^+\right] = 0, \quad \left[b_k^+, b_{k'}^+\right] = 0, \quad \left[P_g^+, P_{g'}^+\right] = 0. \end{gathered} \tag{4.4}$$

We have to remark that the heat bath and the source of energy can contain, besides bosons, fermions too. In the case of Fermi particles we must employ the Hamiltonian which differs from that in Eq. (4.3) (or introduce a new definition of the Fermi operators) and use instead of the commutators $[b_k,\, b^+_{k'}]$ and $[P_g,\, P^+_{g'}]$ in Eq. (4.4) the anticommutators

$$\begin{aligned}\{b_k,\, b^+_{k'}\} &= b_k b^+_{k'} + b^+_{k'} b_k = \delta_{kk'} \\ \{P_g,\, P^+_{g'}\} &= P_g P^+_{g'} + P^+_{g'} P_g = \delta_{gg'}\,.\end{aligned} \tag{4.5}$$

The rate of change of the occupation number in the i-th mode may be determined from the equation of motion (in the Heisenberg representation)

$$\dot{n}_i = -\frac{\mathrm{i}}{\hbar}[n_i,\, H] = -\frac{\mathrm{i}}{\hbar}\{[n_i,\, H_0] + [n_i,\, H_1]\} \tag{4.6}$$

where $n_i = a^+_i a_i$. The commutator $[n_i,\, H_0]$ in Eq. (4.6) contains terms of the type

$$a^+_i a_i a^+_\ell a_\ell - a^+_\ell a_\ell a^+_i a_i = 0$$

and, therefore,

$$[n_i,\, H_0] = 0\,. \tag{4.7}$$

The commutator $[n_i,\, H_1]$ in Eq. (4.6) may be evaluated using relations of the type

$$\begin{aligned}\sum_\ell \sum_k [a^+_i a_i,\, a_\ell b^+_k] &= \sum_\ell \sum_k (a^+_i a_i a_\ell b^+_k - a_\ell b^+_k a^+_i a_i) \\ &= -\sum_\ell \sum_k \delta_{i\ell} a_i b^+_k = -\sum_k a_i b^+_k\end{aligned} \tag{4.8}$$

for the φ term,

$$\begin{aligned}\sum_g \sum_\ell [a^+_i a_i,\, P_g a^+_\ell] &= \sum_g \sum_\ell (a^+_i a_i P_g a^+_\ell - P_g a^+_\ell a^+_i a_i) \\ &= \sum_g \sum_\ell \delta_{i\ell} P_g a^+_i = \sum_g P_g a^+_i\end{aligned} \tag{4.9}$$

for the ζ term, and

$$\begin{aligned}\sum_j \sum_\ell \sum_k [a^+_i a_i,\, a^+_j a_\ell b^+_k] &= \sum_j \sum_\ell \sum_k (a^+_i a_i a^+_j a_\ell b^+_k - a^+_j a_\ell b^+_k a^+_i a_i) \\ &= -\sum_j \sum_\ell \sum_k (\delta_{i\ell} a^+_j a_i b^+_k - \delta_{ij} a^+_i a_\ell b^+_k\ ) \\ &= -\sum_j \sum_k (a^+_j a_i b^+_k - a^+_i a_j b^+_k)\end{aligned} \tag{4.10}$$

for the κ term. The commutation relations given by Eq. (4.4) were used to derive Eqs (4.8)–(4.10). Using relations (4.7)–(4.10) one gets from Eq. (4.6)

$$\dot{n}_i = -\frac{\mathrm{i}}{\hbar}\left\{\sum_k \left(-\varphi a_i b_k^+ + \varphi^* a_i^+ b_k\right) + \sum_g \left(\zeta P_g a_i^+ - \zeta^* P_g^+ a_i\right)\right.$$
$$\left. + \sum_j \sum_k \left[\kappa\left(a_i^+ a_j b_k^+ - a_j^+ a_i b_k^+\right) + \kappa^*\left(a_i^+ a_j b_k - a_j^+ a_i b_k\right)\right]\right\} . \tag{4.11}$$

We will determine the expectation value

$$\langle \dot{n}_i \rangle = \langle \Psi_{\mathrm{I}}(t) | (\dot{n}_i)_{\mathrm{I}} | \Psi_{\mathrm{I}}(t) \rangle$$

where the operator $(\dot{n}_i)_{\mathrm{I}}$ and the state vectors $\langle \Psi_{\mathrm{I}}(t)|$, $|\Psi_{\mathrm{I}}(t)\rangle$ are given in the interaction representation (Appendix 13.1). With the aid of the relations (Appendix 13.2)

$$a_j \exp\left(-\frac{\mathrm{i}}{\hbar} H_0 t\right) = \exp\left(-\frac{\mathrm{i}}{\hbar} H_0 t\right) \exp\left(-\mathrm{i}\omega_j t\right) a_j$$
$$a_j^+ \exp\left(-\frac{\mathrm{i}}{\hbar} H_0 t\right) = \exp\left(-\frac{\mathrm{i}}{\hbar} H_0 t\right) \exp\left(\mathrm{i}\omega_j t\right) a_j^+$$

we obtain the operator $(\dot{n}_i)_{\mathrm{I}}$ in the interaction representation in the form

$$\begin{aligned}
(\dot{n}_i)_{\mathrm{I}} &= \exp\left(\frac{\mathrm{i}}{\hbar} H_0 t\right) \dot{n}_i \exp\left(-\frac{\mathrm{i}}{\hbar} H_0 t\right) \\
&= -\frac{\mathrm{i}}{\hbar} \sum_k \left\{-\varphi a_i b_k^+ \exp\left[\mathrm{i}(-\omega_i + \Omega_k)t\right] + \varphi^* a_i^+ b_k \exp\left[\mathrm{i}(\omega_i - \Omega_k)t\right]\right\} \\
&\quad -\frac{\mathrm{i}}{\hbar} \sum_g \left\{\zeta P_g a_i^+ \exp\left[\mathrm{i}(\omega_i - \Theta_g)t\right] - \zeta^* P_g^+ a_i \exp\left[\mathrm{i}(-\omega_i + \Theta_g)t\right]\right\} \\
&\quad -\frac{\mathrm{i}}{\hbar} \sum_j \sum_k \kappa \left\{a_i^+ a_j b_k^+ \exp\left[\mathrm{i}(\omega_i - \omega_j + \Omega_k)t\right]\right. \\
&\qquad \left. - a_j^+ a_i b_k^+ \exp\left[\mathrm{i}(-\omega_i + \omega_j + \Omega_k)t\right]\right\} \\
&\quad -\frac{i}{\hbar} \sum_j \sum_k \kappa^* \left\{a_i^+ a_j b_k \exp\left[\mathrm{i}(\omega_i - \omega_j - \Omega_k)t\right]\right. \\
&\qquad \left. - a_j^+ a_i b_k \exp\left[\mathrm{i}(-\omega_i + \omega_j - \Omega_k)t\right]\right\} .
\end{aligned} \tag{4.12}$$

We have to determine the state vectors $\langle \Psi_{\mathrm{I}}(t)|$ and $|\Psi_{\mathrm{I}}(t)\rangle$. First of all the interaction Hamiltonian H_{I} has to be evaluated using the adiabatic hypothesis

(Appendix 13.3). We will use the relation

$$H_\mathrm{I} = \exp\left(\frac{\mathrm{i}}{\hbar}H_0 t\right) H_1 \exp\left(-\frac{\mathrm{i}}{\hbar}H_0 t\right) \exp(st) \tag{4.13}$$

where $s \ll 1$ and $\exp(st) \approx 1$ for a finite t. (The term $\exp(st)$ in Eq. (4.13) may be written in the form $\exp(-s|t|)$ too.) From Eq. (4.13) and Eq. (4.3) we get

$$\begin{aligned} H_\mathrm{I} = \exp(st) \Bigg\{ & \sum_j \sum_k \{\varphi a_j b_k^+ \exp[\mathrm{i}(-\omega_j + \Omega_k)t] + \varphi^* a_j^+ b_k \exp[\mathrm{i}(\omega_j - \Omega_k)t]\} \\ & + \sum_j \sum_g \{\zeta P_g a_j^+ \exp[\mathrm{i}(\omega_j - \Theta_g)t] + \zeta^* P_g^+ a_j \exp[\mathrm{i}(-\omega_j + \Theta_g)t]\} \\ & + \sum_j \sum_\ell \sum_k \{\kappa a_j^+ a_\ell b_k^+ \exp[\mathrm{i}(\omega_j - \omega_\ell + \Omega_k)t] \\ & \qquad + \kappa^* a_j a_\ell^+ b_k \exp[\mathrm{i}(-\omega_j + \omega_\ell - \Omega_k)t]\} \Bigg\} . \end{aligned} \tag{4.14}$$

We will use the operator of dynamic evolution in the first approximation in the form (Appendix 13.3)

$$U(t, -\infty) = 1 - \frac{\mathrm{i}}{\hbar} \int_{-\infty}^{t} H_\mathrm{I}(t_1)\mathrm{d}t_1$$

and determine the perturbed state vector $|\Psi_\mathrm{I}(t)\rangle$ in the form

$$\begin{aligned} |\Psi_\mathrm{I}(t)\rangle &= U(t, -\infty)|\Psi_0\rangle \\ &= \Bigg\{ 1 - \frac{1}{\hbar} \sum_j \sum_k \left\{ \varphi a_j b_k^+ \frac{\exp[\mathrm{i}(-\omega_j + \Omega_k)t]}{(-\omega_j + \Omega_k) - \mathrm{i}s} + \varphi^* a_j^+ b_k \frac{\exp[\mathrm{i}(\omega_j - \Omega_k)t]}{(\omega_j - \Omega_k) - \mathrm{i}s} \right\} \\ & \quad - \frac{1}{\hbar} \sum_j \sum_g \left\{ \zeta P_g a_j^+ \frac{\exp[\mathrm{i}(\omega_j - \Theta_g)t]}{(\omega_j - \Theta_g) - \mathrm{i}s} + \zeta^* P_g^+ a_j \frac{\exp[\mathrm{i}(-\omega_j + \Theta_g)t]}{(-\omega_j + \Theta_g) - \mathrm{i}s} \right\} \\ & \quad - \frac{1}{\hbar} \sum_j \sum_\ell \sum_k \left\{ \kappa a_j^+ a_\ell b_k^+ \frac{\exp[\mathrm{i}(\omega_j - \omega_\ell + \Omega_k)t]}{(\omega_j - \omega_\ell + \Omega_k) - \mathrm{i}s} \right. \\ & \qquad \left. + \kappa^* a_j a_\ell^+ b_k \frac{\exp[\mathrm{i}(-\omega_j + \omega_\ell - \Omega_k)t]}{(-\omega_j + \omega_\ell - \Omega_k) - \mathrm{i}s} \right\} \Bigg\} |\Psi_0\rangle \end{aligned} \tag{4.15}$$

where $|\Psi_0\rangle$ is the unperturbed state vector. We assume that the modes in the heat bath as well as the modes in the source of energy have a quasi-continuous distribution. We change summation over Ω_k (and over Θ_g) to integration and use the theory of residues and appropriate contours of integration (Appendix 13.4). As $s \to 0$ we have to use the principal values. We get

$$|\Psi_{\mathrm{I}}(t)\rangle = \left\{1 - \pi\frac{\mathrm{i}}{\hbar}\left[\sum_j (\varphi a_j b_j^+ + \varphi^* a_j^+ b_j) + \sum_j (\zeta P_j a_j^+ + \zeta^* P_j^+ a_j) + \sum_j\sum_\ell \left(\kappa a_j^+ a_\ell b_{\ell-j}^+ + \kappa^* a_j a_\ell^+ b_{j-\ell}\right)\right]\right\} |\Psi_0\rangle \,. \quad (4.16)$$

Likewise, we determine the state vector $\langle \Psi_{\mathrm{I}}(t)|$. The expectation value of the rate of change of the occupation number of the mode with the angular frequency ω_i is given by

$$\langle \Psi_{\mathrm{I}}(t)|(\dot{n}_i)_{\mathrm{I}}|\Psi_{\mathrm{I}}(t)\rangle = \langle \Psi_0|U^*(t,-\infty)(\dot{n}_i)_{\mathrm{I}}U(t,-\infty)|\Psi_0\rangle$$

$$= \langle \Psi_0| \left\{1 + \pi\frac{\mathrm{i}}{\hbar}\left[\sum_m (\varphi^* a_m^+ b_m + \varphi a_m b_m^+) + \sum_m (\zeta^* P_m^+ a_m + \zeta P_m a_m^+) + \sum_m\sum_\ell (\kappa^* a_m a_\ell^+ b_{m-\ell} + \kappa a_m^+ a_\ell b_{\ell-m}^+)\right]\right\} \times$$

$$\left(-\frac{\mathrm{i}}{\hbar}\right)\left\{\sum_k \{-\varphi a_i b_k^+ \exp[\mathrm{i}(-\omega_i + \Omega_k)t] + \varphi^* a_i^+ b_k \exp[\mathrm{i}(\omega_i - \Omega_k)t]\}\right.$$
$$+ \sum_g \{\zeta P_g a_i^+ \exp[\mathrm{i}(\omega_i - \Theta_g)t] - \zeta^* P_g^+ a_i \exp[\mathrm{i}(-\omega_i + \Theta_g)t]\}$$
$$+ \sum_j\sum_k \kappa \{a_i^+ a_j b_k^+ \exp[\mathrm{i}(\omega_i - \omega_j + \Omega_k)t]$$
$$- a_j^+ a_i b_k^+ \exp[\mathrm{i}(-\omega_i + \omega_j + \Omega_k)t]\}$$
$$+ \sum_j\sum_k \kappa^* \{a_i^+ a_j b_k \exp[\mathrm{i}(\omega_i - \omega_j - \Omega_k)t]$$
$$\left. - a_j^+ a_i b_k \exp[\mathrm{i}(-\omega_i + \omega_j - \Omega_k)t]\}\right\} \times$$

$$\left\{1 - \pi\frac{\mathrm{i}}{\hbar}\left[\sum_{m'} (\varphi a_{m'} b_{m'}^+ + \varphi^* a_{m'}^+ b_{m'}) + \sum_{m'} (\zeta P_{m'} a_{m'}^+ + \zeta^* P_{m'}^+ a_{m'}) + \sum_{m'}\sum_\ell (\kappa a_{m'}^+ a_\ell b_{\ell-m'}^+ + \kappa^* a_{m'} a_\ell^+ b_{m'-\ell})\right]\right\} |\Psi_0\rangle \,. \quad (4.17)$$

The operators in Eq. (4.17) are Hermitian. The eigenfunctions (the eigen state vectors) of a Hermitian operator associated with different eigenvalues are orthogonal. Therefore, only the terms with operators that do not alter the corresponding bra and ket state vectors can contribute to the result. These terms determine the expectation values of the occupation numbers. We select the nonvanishing terms from Eq. (4.17). We will replace the expectation value of a product by the product of expectation values. The φ terms yield

$$\begin{aligned}\langle \dot{n}_i \rangle_{\Phi} &= \frac{\pi}{\hbar^2}\varphi^*\varphi\langle \Psi_0 | \sum_m \sum_k \{a_m b_m^+ a_i^+ b_k \exp\left[\mathrm{i}(\omega_i - \Omega_k)t\right] \\ &\quad + a_i b_k^+ a_m^+ b_m \exp\left[\mathrm{i}(-\omega_i + \Omega_k)t\right] - a_m^+ b_m a_i b_k^+ \exp\left[\mathrm{i}(-\omega_i + \Omega_k)t\right] \\ &\quad - a_i^+ b_k a_m b_m^+ \exp\left[\mathrm{i}(\omega_i - \Omega_k)t\right]\} \,|\, \Psi_0\rangle \\ &= -\frac{2\pi}{\hbar^2}\varphi^*\varphi\langle a_i^+ a_i (b_i^+ b_i + 1) - (a_i^+ a_i + 1) b_i^+ b_i \rangle \\ &= -\frac{2\pi}{\hbar^2}\varphi^*\varphi\left[\langle n_i\rangle(\langle N_i\rangle + 1) - (\langle n_i\rangle + 1)\langle N_i\rangle\right] \\ &= -\Phi_i\left[\langle n_i\rangle \exp\left(\frac{\hbar\omega_i}{kT}\right) - (\langle n_i\rangle + 1)\right] \end{aligned} \tag{4.18}$$

where

$$\begin{aligned}\Phi_i &= \frac{2\pi}{\hbar^2}|\varphi|^2\langle N_i\rangle \\ \langle N_i\rangle &= \langle \Psi_0 | b_i^+ b_i | \Psi_0\rangle = \frac{1}{\exp\left(\dfrac{\hbar\omega_i}{kT}\right) - 1}\end{aligned} \tag{4.19}$$

is the transition probability of the energy transfer between the vibration system and the heat bath and the occupation number of the vibration mode in the heat bath (which is in thermodynamic equilibrium), respectively. $\langle n_i\rangle$ is the expectation value of the occupation number of the vibration mode i.

Likewise, the ζ terms yield the result

$$\langle s_i\rangle = \frac{2\pi}{\hbar^2}|\zeta|^2\langle P_i^+ P_i - a_i^+ a_i\rangle = \frac{2\pi}{\hbar^2}|\zeta|^2\left(\langle S_i\rangle - \langle n_i\rangle\right) \tag{4.20}$$

where $\langle S_i\rangle = \langle P_i^+ P_i\rangle$ is the occupation number of the mode with frequency ω_i in the source of energy and s_i is the number of quanta supplied to the vibration mode of the same frequency. If $\langle S_i\rangle \gg \langle n_i\rangle$ then Eq. (4.20) yields approximately

$$\langle s_i\rangle = \frac{2\pi}{\hbar^2}|\zeta|^2\langle S_i\rangle\,.$$

To evaluate the κ terms we separate the terms for $\omega_j > \omega_i$ and for $\omega_j < \omega_i$. For $\omega_j > \omega_i$ we get

$$\begin{aligned}\langle \dot{n}_i \rangle_\chi = &\frac{\pi}{\hbar^2}|\kappa|^2 \langle \Psi_0 | \sum_\ell \sum_m \sum_j \sum_k \{ a_m a_\ell^+ b_{m-\ell} a_i^+ a_j b_k^+ \exp[\mathrm{i}(\omega_i - \omega_j + \Omega_k)t] \\ &+ a_j^+ a_i b_k a_m^+ a_\ell b_{\ell-m}^+ \exp[\mathrm{i}(-\omega_i + \omega_j - \Omega_k)t] \\ &- a_m^+ a_\ell b_{\ell-m}^+ a_j^+ a_i b_k \exp[\mathrm{i}(-\omega_i + \omega_j - \Omega_k)t] \\ &- a_i^+ a_j b_k^+ a_m a_\ell^+ b_{m-\ell} \exp[\mathrm{i}(\omega_i - \omega_j + \Omega_k)t] \} \,|\Psi_0\rangle \\ = &-\frac{\pi}{\hbar^2}|\kappa|^2 \left\langle \sum_j \{ a_i^+ a_i (a_j^+ a_j + 1) b_{ji}^+ b_{ji} - (a_i^+ a_i + 1) a_j^+ a_j (b_{ji}^+ b_{ji} + 1) \} \right\rangle \\ = &-\frac{\pi}{\hbar^2}|\kappa|^2 \sum_j \{ \langle n_i \rangle (\langle n_j \rangle + 1) \langle N_{j-i} \rangle - (\langle n_i \rangle + 1) \langle n_j \rangle (\langle N_{j-i} \rangle + 1) \} \\ = &-\sum_j \chi_{ij} \left\{ \langle n_i \rangle (\langle n_j \rangle + 1) \exp\left(\frac{\hbar\omega_i}{kT}\right) - (\langle n_i \rangle + 1) \langle n_j \rangle \exp\left(\frac{\hbar\omega_j}{kT}\right) \right\}\end{aligned} \tag{4.21}$$

where

$$\begin{aligned}\chi_{ij} &= \frac{\pi}{\hbar^2}|\kappa|^2 \left(\langle N_{j-i} \rangle + 1\right) \exp\left(-\frac{\hbar\omega_j}{kT}\right) \\ \langle N_{j-i} \rangle &= \frac{1}{\exp\left[\dfrac{\hbar(\omega_j - \omega_i)}{kT}\right] - 1} .\end{aligned} \tag{4.22}$$

For $\omega_j < \omega_i$ we get a similar result but

$$\begin{aligned}\chi_{ij} &= \frac{\pi}{\hbar^2}|\kappa|^2 \langle N_{i-j} \rangle \exp\left(-\frac{\hbar\omega_j}{kT}\right) \\ \langle N_{i-j} \rangle &= \frac{1}{\exp\left[\dfrac{\hbar(\omega_i - \omega_j}{kT}\right] - 1} .\end{aligned} \tag{4.23}$$

N_{j-i} and N_{i-j} are the occupation numbers in the vibration modes in the heat bath with the frequencies $\omega_j - \omega_i$ and $\omega_i - \omega_j$, respectively. In Eq. (4.21) we summed over k to evaluate the expectation value; on account of the principle of the conservation of momentum two terms are nonvanishing for the positive value of k ($k = \ell - m > 0$), the others for the negative value of k ($k = m - \ell < 0$).

Eqs (4.18), (4.20), (4.21), and (4.23) enable us to write the rate equation for the expectation value of the occupation number of the vibration mode with the frequency ω_i in the form

$$\langle \dot{n}_i \rangle = \langle s_i \rangle + \langle \dot{n}_i \rangle_\Phi + \langle \dot{n}_i \rangle_\chi = \langle s_i \rangle - \Phi_i \left[\langle n_i \rangle \exp\left(\frac{\hbar\omega_i}{kT}\right) - (\langle n_i \rangle + 1) \right]$$
$$- \sum_j \chi_{ij} \left[\langle n_i \rangle (\langle n_j \rangle + 1) \exp\left(\frac{\hbar\omega_i}{kT}\right) - (\langle n_i \rangle + 1) \langle n_j \rangle \exp\left(\frac{\hbar\omega_j}{kT}\right) \right] . \tag{4.24}$$

This equation is precisly the same as that postulated by Fröhlich. The χ term in Eq. (4.24) stands for two quanta processes. The χ transition probability was assessed in [101] from the relaxation time ($\chi = 4 \times 10^{-2}\,\mathrm{s}^{-1}$). The microscopic technique makes possible to determine the Fröhlich rate equation as an equation of expectation values.

4.2 Validity of the Microscopic Foundation of the Rate Equation

In the previous section the Fröhlich rate equation was derived from the Wu–Austin Hamiltonian using perturbation theory. Some authors tried to prove that the Wu–Austin Hamiltonian cannot be used, that it does not make basis for derivation of the Fröhlich rate equation, and, therefore, that the Fröhlich rate equation does not have microscopic foundation, e.g. in [102,103]. They argued that in this case perturbation theory does not yield a correct result. For instance, they found solution derived by a special transformation technique with the conclusion that the Wu–Austin Hamiltonian leads to energy condensation at all temperatures even with no energy supply from external sources. We will assess these objections.

A special case was analyzed in [102,103] based on the assumption that the frequencies of all the normal modes of the vibration system are the same, i.e. $\omega_i = \omega_j = \omega$ for $i \neq j$. This "degenerate" one frequency vibration system does not correspond to a real physical vibration system. Interaction alters the frequencies of vibrations. For instance, if two identical harmonic oscillators interact their frequency is changed and they simultaneously oscillate at two frequencies – one is higher and the other is lower than the original frequency. The frequency shift depends on the strength of interaction. We have to point out that Fröhlich assumed vibration modes in the energy band of the ΔE width and not vibrations with one discrete energy. In [102,103] the Wu–Austin Hamiltonian given by Eqs (4.1)–(4.3) was evaluated under this condition of energy "degeneration".

The last term on the right hand side of the perturbation Hamiltonian H_1

given by Eq. (4.3)

$$\sum_j \sum_i \sum_k \left(\kappa a_j^+ a_i b_k^+ + \kappa^* a_j a_i^+ b_k\right) \tag{4.25}$$

stands for the energy transfer between two vibration modes with the participation of the heat bath and causes energy condensation in the system. This term describing the microprocesses of the energy transfer between vibration modes is fully justified as far as the heat bath balances the differences between the energies and momentums of the interacting modes. The term given by Eq. (4.25) is used in [102,103] too. On account of the principle of the conservation of energy and momentum only the terms for $\omega_i - \omega_j = \Omega_k$, $i - j = k$ will contribute (where Ω_k, k represent the frequency and the momentum of the heat bath modes, respectively). For $\omega_i = \omega_j$ the operator a_i^+ (or a_j^+) creates and the operator a_i (or a_j) annihilates the same quantum of energy and no energy is transferred to or from the heat bath. The condition $\omega_i = \omega_j$ states that only the case $\Omega_k = 0$ is taken into account in [102] but in this case the mathematical operations should be modified to avoid the term $1/\Omega_k$, which is in one of the final relations derived in [102]

$$-N_0^2 M^2 |\chi|^2 \sum_k \frac{1}{4\Omega_k} .$$

Real physical vibration systems have an energy band of certain width and only several vibration modes may have the same energy, not all modes. We may conclude that the term given by Eq. (4.25) does not describe real physical microprocesses if it is assumed that the frequencies of all the vibration modes are equal.

Only momentum may be transferred between the vibration modes in the vibration system and the vibration modes in the heat bath if $\omega_i = \omega_j = \omega$ and $i \neq j$. Even in the systems with translation symmetry several modes may have the same frequency in the first Brillouin zone (the frequency is determined by the dispersion relation $\omega = \omega(k)$). On account of the conservation of momentum the wave vectors should satisfy special relations to enable momentum transfer which is rather an extraordinary case.

We conclude that the result "the sytem can lower its energy by just increasing the number of quanta N_0" ($N_0 \to \infty$ for no energy supply) in [102,103] is derived from the Hamiltonian under the conditions that contradict the properties of the real Fröhlich system.

In connection with the results given in [102,103] we may discuss the question of the energy of the ground state of the Wu–Austin Hamiltonian. We will employ a classical analogue. If the energy is given by the relation $E = ax^2 + bx^3$ (where $a > 0,\ b > 0$) then for $x \to -\infty$ the energy of the system tends to $-\infty$. Nevertheless, even a small term cx^4 (where $0 < c \ll 1$) that has a negligible effect in the "working" region of x can provide reasonable behaviour of the system for $x \to -\infty$.

4.3 Derivation of the Fröhlich Rate Equation Using Green Function Technique

We will study the problem of the Fröhlich rate equation from a microscopic point of view using the finite temperature Green function method including all orders of interaction strength.

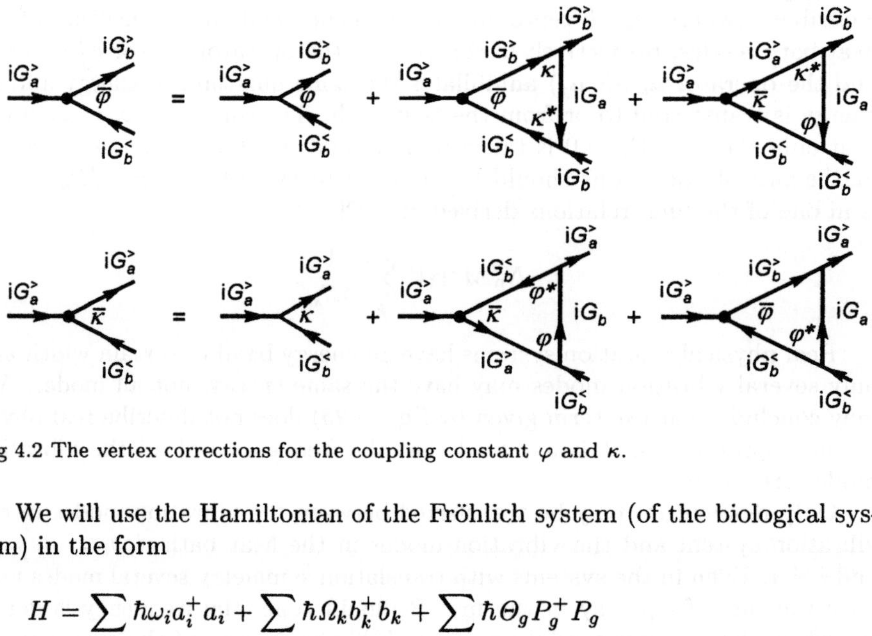

Fig 4.2 The vertex corrections for the coupling constant φ and κ.

We will use the Hamiltonian of the Fröhlich system (of the biological system) in the form

$$H = \sum_i \hbar\omega_i a_i^+ a_i + \sum_k \hbar\Omega_k b_k^+ b_k + \sum_g \hbar\Theta_g P_g^+ P_g$$
$$+\frac{1}{2}\sum_{i,j,k}\left(\varphi b_j b_k^+ a_i + \varphi^* b_j^+ b_k a_i^+\right) + \frac{1}{2}\sum_{i,j,k}\left(\kappa a_j^+ a_i b_k^+ + \kappa^* a_j a_i^+ b_k\right)$$
$$+\sum_{i,g}\left(\zeta P_g a_i^+ + \zeta^* P_g^+ a_i\right) \tag{4.26}$$

where $\sum_{i,j,k}$ means summation over i, j, and k; φ, κ and ζ are the coupling constants for the heat bath (the one quantum process, the two quanta process), and the source of energy with the vibration modes, respectively. Strictly speaking, we should consider higher order and possibly anharmonic terms; however, it can be shown for many systems that accounting for those terms have only minor effects and, within certain limits, they may be ignored altogether. The Hamiltonian given by Eq. (4.26) is slightly different from the "historical" Hamiltonian given by Eqs (4.1)–(4.3) but physically more precise. Nevertheless, we will show that the final results – i.e. the rate equations derived from both these Hamiltonians – correspond one another.

The electromagnetic oscillations are bosons. The excitations within the heat bath and the energy source can be either bosons, or fermions, e.g. phonons or electrons. The operators b_k^+, b_k, P_g^+, P_g satisfy the commutation or anticommutation relations given by Eqs (4.4) and (4.5). In the case of fermions b_k^+, b_k, P_g^+, P_g are not the ordinary operators. These operators create or annihilate an excited state of fermion in k state (g state). They represent transition of fermion from the ground state to the excited state and vice versa (e.g. electron is transferred from the bound state to conduction band). This is a very important process in biological systems.

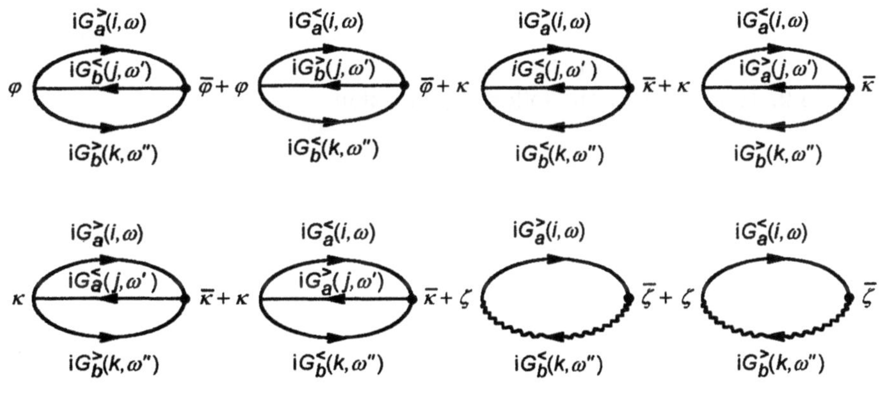

Fig. 4.3 Diagrams represent the rate of change of the number of phonons with energy ω_i.

We will use the interaction picture. The rate of change of the occupation number in the i-th mode is given by the relation

$$\dot{n}_i = -\frac{\mathrm{i}}{\hbar}[n_i, H] = -\frac{\mathrm{i}}{\hbar}(n_i H - H n_i) \tag{4.27}$$

with $n_i = a_i^+ a_i$; the occupation number in the i-th mode.

The expectation value of the rate of change of the number of phonons with energy ω_i in the biological system is given by

$$\langle \dot{n}_i \rangle = \langle \Psi(t)|\dot{n}_i(t)|\Psi(t)\rangle = \frac{2}{\hbar}\Im\left\{\sum_{j,k} \varphi^* \langle \Psi(t)|a_i^+(t) b_j(t) b_k^+(t)|\Psi(t)\rangle\right.$$

$$+\sum_{j,k} \kappa\langle \Psi(t)|a_i^+(t)a_j(t)b_k^+(t)|\Psi(t)\rangle + \sum_{j,k} \kappa^*\langle \Psi(t)|a_i^+(t)a_j(t)b_k(t)|\Psi(t)\rangle$$

$$\left.+\sum_g \zeta\langle \Psi(t)|a_i^+(t)P_g(t)|\Psi(t)\rangle \right\} \tag{4.28}$$

where $\Im$ denotes the imaginary part of the expression, $|\Psi(t)\rangle$ is the state vector of the whole system, and the angular bracket is the ensemble average. The ensemble average $\langle\hat{O}\rangle$ for any operator $\hat{O}$ is obtained by $\langle\hat{O}\rangle = \mathrm{Tr}(\rho\hat{O})$ where ρ is the density operator defined by

$$\rho = \frac{\exp\left(-\dfrac{H}{kT}\right)}{\mathrm{Tr}\left[\exp\left(-\dfrac{H}{kT}\right)\right]} . \tag{4.29}$$

Using the finite temperature Green function technique [104], one can easily obtain the rate of change of the number of phonons with energy ω_i as shown in Fig. 4.2. It is given by the following equation

$$\langle\dot{n}_i\rangle = -\frac{2\varphi\overline{\varphi}^*}{\hbar^2}\Im\left\{\sum_{j,k}\int\int\int\frac{\mathrm{d}\omega\mathrm{d}\omega'\mathrm{d}\omega''}{(2\pi)^3}\left[G_a^>(i,\omega)G_b^<(j,\omega')G_b^>(k,\omega'')\right.\right.$$

$$\left.\left.+G_a^<(i,\omega)G_b^>(j,\omega')G_b^<(k,\omega'')\right]\delta(\omega-\omega'+\omega'')\right\}$$

$$-\frac{2\kappa\overline{\kappa}^*}{\hbar^2}\Im\left\{\sum_{j,k}\int\int\int\frac{\mathrm{d}\omega\mathrm{d}\omega'\mathrm{d}\omega''}{(2\pi)^3}\left\{\left[G_a^>(i,\omega)G_a^<(j,\omega')G_b^<(k,\omega'')\right.\right.\right.$$

$$\left.+G_a^<(i,\omega)G_a^>(j,\omega')G_b^>(k,\omega'')\right]\delta(\omega-\omega'-\omega'')$$

$$+\left[G_a^>(i,\omega)G_a^<(j,\omega')G_b^>(k,\omega'')\right.$$

$$\left.\left.\left.+G_a^<(i,\omega)G_a^>(j,\omega')G_b^>(k,\omega'')\right]\right\}\delta(\omega-\omega'+\omega'')\right\}$$

$$+\frac{2\zeta\overline{\zeta}^*}{\hbar^2}\Im\left\{\sum_g\int\int\frac{\mathrm{d}\omega\mathrm{d}\omega'}{(2\pi)^2}\left[G_a^>(i,\omega)G_P^<(g,\omega')\right.\right.$$

$$+G_a^{<}(i,\omega)G_P^{>}(g,\omega')\Big]\,\delta(\omega-\omega')\Bigg\} \tag{4.30}$$

where $G_a^{>}(i,\omega)$ ($G_a^{<}(i,\omega)$), $G_b^{>}(i,\omega)$ ($G_b^{<}(i,\omega)$), and $G_P^{>}(i,\omega)$ ($G_P^{\lessgtr}(i,\omega)$) are the retarded and advanced Fourier transforms of the finite temperature Green functions defined by

$$\begin{aligned} G_a(i,t) &= -\mathrm{i}\langle\Psi(t)|T\{a_i(t)a_i^+(0)\}|\Psi(t)\rangle \\ G_b(i,t) &= -\mathrm{i}\langle\Psi(t)|T\{b_i(t)b_i^+(0)\}|\Psi(t)\rangle \\ G_P(i,t) &= -\mathrm{i}\langle\Psi(t)|T\{P_i(t)P_i^+(0)\}|\Psi(t)\rangle \end{aligned} \tag{4.31}$$

where T is the time-ordering operator. The retarded and advanced Fourier transforms of the above defined Green functions can be written as

$$\begin{aligned} G_a^{>}(i,\omega) &= \frac{1+n(\overline{\omega}_i)}{\omega-\overline{\omega}_i+\mathrm{i}\delta} \\ G_a^{<}(i,\omega) &= \frac{-n(\overline{\omega}_i)}{\omega-\overline{\omega}_i-\mathrm{i}\delta}\,. \end{aligned} \tag{4.32}$$

The retarded and advanced functions for the heat bath and the source of energy have a similar form

$$\begin{aligned} G_b^{>}(i,\omega) &= \frac{1+N_b(\overline{\Omega}_i)}{\omega-\overline{\Omega}_i+\mathrm{i}\delta} \\ G_b^{<}(i,\omega) &= \frac{-N_b(\overline{\Omega}_i)}{\omega-\overline{\Omega}_i-\mathrm{i}\delta} \\ G_P^{>}(i,\omega) &= \frac{1+N_P(\overline{\Theta}_i)}{\omega-\overline{\Theta}_i+\mathrm{i}\delta} \\ G_P^{\lessgtr}(i,\omega) &= \frac{-N_P(\overline{\Theta}_i)}{\omega-\overline{\Theta}_i-\mathrm{i}\delta} \end{aligned} \tag{4.33}$$

where $n(\overline{\omega}_i)$, $N_b(\overline{\Omega}_i)$, and $N_P(\overline{\Theta}_i)$ are the quantum distributions of the biological system, the heat bath, and the source of energy, respectively. $\overline{\omega}_i = \omega_i + \sum_a(i)$ is the modified excitation of the biological system phonon with energy ω_i when there is no interaction between the biological system and heat reservoir, and $\sum_a(i)$ is the proper self energy of the excitation. Due to the nature of the weak dipolar interaction, one expects the self energy, $\sum_a(i)$, to be small. In general $\sum_a(i)$ depends on ω_i, but the energy band is very narrow; it can be approximated to be a constant. Similarly, $\overline{\Omega}_i = \Omega_i + \sum_b(i)$

and $\overline{\Theta}_i = \Theta_i + \sum_P(i)$ where $\sum_b(i)$ and $\sum_P(i)$ are the proper self energies for the heat bath and energy supply excitations, respectively. The vertex corrections $\overline{\varphi}$, $\overline{\kappa}$, and $\overline{\zeta}$ consist of all possible proper corrections to all orders of interaction. The actual detail forms of the vertex corrections $\overline{\varphi}$, $\overline{\kappa}$ can be obtained from the coupling equations as shown diagrammatically in Fig. 4.3. The coupling strengths $\overline{\varphi}$, $\overline{\kappa}$, and $\overline{\zeta}$ are very complicated quantities and in general depend on ω_i. Since the energy band of the biological system under this study is very narrow, the coupling strengths and their vertex corrections will be considered to be constants.

Substituting Eqs (4.32) and (4.33) into Eq. (4.30) and integrating over ω, ω', and ω'', one gets

$$\begin{aligned}\langle \dot{n}_i \rangle = &-\frac{\varphi\overline{\varphi}}{\pi\hbar^2}\Im\sum_{j,k}\left\{\frac{n(\overline{\omega}_i)[1+N_b(\overline{\Omega}_j)]N_b(\overline{\Omega}_k)}{\overline{\Omega}_j-\overline{\Omega}_k-\overline{\omega}_i-\mathrm{i}\delta}\right.\\ &\left.+\frac{[1+n(\overline{\omega}_i)]N_b(\overline{\Omega}_j)[1+N_b(\overline{\Omega}_k)]}{\overline{\Omega}_j-\overline{\Omega}_k-\overline{\omega}_i+\mathrm{i}\delta}\right\}\\ &-\frac{\kappa\overline{\kappa}}{\pi\hbar^2}\Im\sum_{j,k}\left\{\frac{n(\overline{\omega}_i)[1+n(\overline{\omega}_j)][1+N_b(\overline{\Omega}_k)]}{\overline{\omega}_j+\overline{\Omega}_k-\overline{\omega}_i-\mathrm{i}\delta}\right.\\ &+\frac{[1+n(\overline{\omega}_i)]n(\overline{\omega}_j)N_b(\overline{\Omega}_k)}{\overline{\omega}_j+\overline{\Omega}_k-\overline{\omega}_i+\mathrm{i}\delta}+\frac{n(\overline{\omega}_i)[1+n(\overline{\omega}_j)]N_b(\overline{\Omega}_k)}{\overline{\omega}_j-\overline{\Omega}_k-\overline{\omega}_i-\mathrm{i}\delta}\\ &\left.+\frac{[1+n(\overline{\omega}_i)]n(\overline{\omega}_j)[1+N_b(\overline{\Omega}_k)]}{\overline{\omega}_j-\overline{\Omega}_k-\overline{\omega}_i+\mathrm{i}\delta}\right\}\\ &-\frac{\zeta\overline{\zeta}}{\pi\hbar^2}\Im\sum_{j}\left\{\frac{n(\overline{\omega}_i)[1+N_P(\overline{\Theta}_j)]}{\overline{\omega}_i-\overline{\Theta}_j-\mathrm{i}\delta}+\frac{[1+n(\overline{\omega}_i)]N_P(\overline{\Theta}_j)]}{\overline{\omega}_i-\overline{\Theta}_j+\mathrm{i}\delta}\right\}.\end{aligned} \tag{4.34}$$

Summing over j and taking the imaginary part, Eq. (4.34) becomes

$$\begin{aligned}\langle \dot{n}_i \rangle = -\frac{\varphi\overline{\varphi}}{\hbar^2}\sum_j\Big\{&n(\overline{\omega}_i)[1+N_b(\overline{\omega}_i+\overline{\Omega}_j)]N_b(\overline{\Omega}_j)\\ &-[1+n(\overline{\omega}_i)]N_b(\overline{\omega}_i+\overline{\Omega}_j)[1+N_b(\overline{\Omega}_j)]\Big\}\end{aligned}$$

$$-\frac{\kappa\overline{\kappa}}{\hbar^2}\sum_j\Big\{n(\overline{\omega}_i)[1+n(\overline{\omega}_j][1+N_b(\overline{\omega}_i-\overline{\omega}_j)]$$
$$-[1+n(\overline{\omega}_i)]n(\overline{\omega}_j)N_b(\overline{\omega}_i-\overline{\omega}_j)]$$
$$+n(\overline{\omega}_i)[1+n(\overline{\omega}_j)]N_b(\overline{\omega}_j-\overline{\omega}_i)$$
$$-[1+n(\overline{\omega}_i)]n(\overline{\omega}_j)[1+N_b(\overline{\omega}_j-\overline{\omega}_i)]\Big\}$$
$$-\frac{\zeta\overline{\zeta}}{\hbar^2}\Big\{n(\overline{\omega}_i)[1+N(\overline{\omega}_i)]-[1+n(\overline{\omega}_i)]N(\overline{\omega}_i)\Big\} \tag{4.35}$$

Since the heat bath is big compared to the biological system under consideration, it is a good approximation to assume that the energy transfer from the vibration system does not have any significant effect on the heat bath which remains in thermodynamic equilibrium. Therefore, the excitation of the heat bath can be neglected and substituting relation

$$N_b(\overline{\omega}_i)=N(\overline{\omega}_i)=[\exp\left(\frac{\hbar\overline{\omega}_i}{kT}\right)\pm 1]^{-1}$$

in Eq. (4.35) yields

$$\langle\dot{n}_i\rangle = s_i-\Phi_i(T,\overline{\omega}_i)\left\{n(\overline{\omega}_i)\exp\left(\frac{\hbar\overline{\omega}_i}{kT}\right)-[n(\overline{\omega}_i)+1]\right\}$$
$$-\sum_j\chi_{ij}(T,\overline{\omega}_i,\overline{\omega}_j)\Big\{n(\overline{\omega}_i)[n(\overline{\omega}_j)+1]\exp\left(\frac{\hbar\overline{\omega}_i}{kT}\right)$$
$$-n(\overline{\omega}_j)[n(\overline{\omega}_i)+1]\exp\left(\frac{\hbar\overline{\omega}_j}{kT}\right)\Big\} \tag{4.36}$$

where

$$s_i=\pi\frac{\zeta\overline{\zeta}}{\hbar^2}\left[N_P(\overline{\omega}_i)-n(\overline{\omega}_i)\right] \tag{4.37}$$
$$\Phi_i(T,\overline{\omega}_i)=\pi\frac{\varphi\overline{\varphi}}{\hbar^2}\left[1\pm N(\overline{\omega}_i)\right]\exp\left(-\frac{\hbar\overline{\omega}_i}{kT}\right) \tag{4.38}$$

and

$$\chi_i(T,\overline{\omega}_i,\overline{\omega}_j)=\frac{\pi\kappa\overline{\kappa}}{2\hbar^2}\exp\left(-\frac{\hbar\overline{\omega}_j}{kT}\right)\begin{matrix}\nearrow & 1\pm N(\overline{\omega}_j-\overline{\omega}_i)\,, & \overline{\omega}_j>\overline{\omega}_i\\ \searrow & N(\overline{\omega}_i-\overline{\omega}_j)\,, & \overline{\omega}_j<\overline{\omega}_i\end{matrix}\ . \tag{4.39}$$

One notices that s_i given in Eq. (4.37) is the net power absorbed by the i-th mode. Eq. (4.36) is exactly the same rate equation suggested by Fröhlich.

The Hamiltonian in Eqs (4.1)–(4.3) and in Eq. (4.26) contains only those terms that are assumed to have the major effects on behaviour of the system. The terms with minor effects are neglected. Livshits [105,106] included a higher order term and tried to prove its strong effect on behaviour of the system. He assumed that processes of simultaneous emission–emission and absorption–absorption in two vibration modes with different frequencies are as important as the Fröhlich processes absorption–emission. We will not assess here the probability of these processes in real systems. We will derive the Livshits rate equation. The Hamiltonian in Eq. (4.26) can be modified to take care of the contributions from creating two quanta or destroying two quanta simultaneously. Thus,

$$\begin{aligned} H = & \sum_i \hbar\omega_i a_i^+ a_i + \sum_k \hbar\Omega_k b_k^+ b_k + \sum_g \hbar\Theta_g P_g^+ P_g \\ & + \frac{1}{2}\sum_{i,j,k} (\varphi b_j b_k^+ a_i + \varphi^* b_j^+ b_k a_i^+) + \frac{1}{2}\sum_{i,j,k} (\kappa a_j^+ a_i b_k^+ + \kappa^* a_j a_i^+ b_k) \\ & + \frac{1}{2}\sum_{i,j,k} (\xi a_j a_i b_k^+ + \xi^* a_j^+ a_i^+ b_k) + \sum_{i,g} (\zeta P_g a_i^+ + \zeta^* P_g^+ a_i) \end{aligned} \tag{4.40}$$

where ξ is the coupling constant for the simultaneous creation of two quanta or destruction of two quanta. Using this modified Hamiltonian and repeating the same calculation for $\langle \dot{n} \rangle$, we get

$$\begin{aligned} \langle \dot{n}_i \rangle = & \ s_i - \Phi_i \left[\langle n_i \rangle \exp\left(\frac{\hbar\omega_i}{kT}\right) - (\langle n_i \rangle + 1) \right] \\ & - \sum_j \chi_{ij} \left[\langle n_i \rangle (\langle n_j \rangle + 1) \exp\left(\frac{\hbar\omega_i}{kT}\right) - (\langle n_i \rangle + 1) \langle n_j \rangle \exp\left(\frac{\hbar\omega_j}{kT}\right) \right] \\ & - \sum_j \Xi_{ij} \left\{ \langle n_i \rangle \langle n_j \rangle \exp\left[\frac{\hbar(\omega_i + \omega_j)}{kT}\right] - (\langle n_i \rangle + 1)(\langle n_j \rangle + 1) \right\} \end{aligned} \tag{4.41}$$

where s_i, Φ_i, and χ_{ij} are the same as in Eqs (4.37), (4.38), (4.39), and

$$\Xi_{ij} = \frac{\pi |\xi|^2}{\hbar^2} N(\omega_i + \omega_j) \,. \tag{4.42}$$

The Livshits rate equation given by (4.41) contains additional interaction terms which can affect energy condensation in the Fröhlich systems as will be analyzed in Chapter 5.

The rate equation in this section includes multiple quanta processes. Using the Green function method and the vertex correction as shown in the Figs 4.2 and 4.3 we have calculated to the infinite order of interactions.

We use the effective coupling constants $\overline{\varphi}$, $\overline{\kappa}$ and normalized energies $\overline{\omega}$, $\overline{\Omega}$ for all excitations both in the system and in the heat bath. If we take the first order approximation the result in this section will reduce to that of the perturbation result in the previous section.

4.4 Non-Linear Rate Equation

Fröhlich assumed two quanta interactions of a special kind: energy transfer between two modes of the vibration system with participation of the heat bath. The vibration system may be linear but the interaction with the heat bath introduces non-linear effects and enable two quanta processes. Nevertheless, the vibration system may have non-linear properties by themselves and the multiple quanta transfer may exist without the Fröhlich interaction with the heat bath [107,108]. We will not analyze here the concrete non-linear mechanisms in biological macromolecules and structures. We note that strong electric fields in proteins, membranes, and cytoskeleton polymers can be responsible for the non-linearity. In the preceding sections we assumed that the elastic force acting on the oscillating particles is proportional to the displacement from the equilibrium position. Nonetheless, the dependence of the elastic force on the displacement can be non-linear which seems to be a quite general phenomenon especially in folded chains containing particles of different types. The non-linear mechanisms enable two quanta and multiple quanta transfer between vibration modes. A non-linear system with multiple quanta transfer may be described by the Hamiltonian

$$H = H_0 + H_1' + H_2 \tag{4.43}$$

where H_0 is given by Eq. (4.2) and H_1', H_2 are given by

$$H_1' = \sum_i \sum_k (\varphi a_i b_k^+ + \varphi^* a_i^+ b_k) + \sum_i \sum_g (\zeta P_g a_i^+ + \zeta^* P_g^+ a_i) \tag{4.44}$$

$$\begin{aligned} H_2 = \sum_{j,\ell,p} \sum_{m,n} \{ & [\sigma a_j^+ (a_\ell)^m (a_p)^n + \sigma^* a_j (a_\ell^+)^m (a_p^+)^n] \\ & + [\varrho a_j^+ (a_\ell^+)^m (a_p)^n + \varrho^* a_j (a_\ell)^m (a_p^+)^n] \\ & + [\xi a_j^+ (a_\ell)^m (a_p^+)^n + \xi^* a_j (a_\ell^+)^m (a_p)^n] \} \ . \end{aligned} \tag{4.45}$$

The perturbation Hamiltonian has two terms – H_1' and H_2 given by Eq. (4.44) and Eq. (4.45), respectively. The Hamiltonian H_1' contains the first two perturbation terms given on the right hand side of Eq. (4.3), i.e. the φ and ζ terms which stand for the linear coupling to the heat bath and the linear coupling to the source of energy, respectively. We did not include the Fröhlich

χ term and, therefore, we neglected the two quanta processes of interaction of the vibration system with the heat bath. In Eq. (4.45) σ, ϱ, ξ are the coupling constants and symbols of the type $(a_i)^m$ stand for $a_i a_i a_i \ldots a_i$ m-times (and $(a_i)^m = 1$ for $m = 0$).

We will use the equation of motion in the form

$$\dot{n}_i = -\frac{\mathrm{i}}{\hbar}\,[n_i,\, H] = -\frac{\mathrm{i}}{\hbar}\,\{[n_i,\, H_0] + [n_i,\, H_1'] + [n_i,\, H_2]\} \tag{4.46}$$

where $[n_i,\, H_0] = 0$ as is given by Eq. (4.7). $[n_i,\, H_1']$ yields the same result as is given by Eqs (4.18) and (4.20) for the φ and ζ terms of the Hamiltonian H_1, respectively. We will analyze the commutator $[n_i,\, H_2]$ in Eq. (4.46). To avoid complicated and incomprehensive formulas we will evaluate separately the first two terms from the right hand side of Eq. (4.45), i.e. the σ terms. This part of the Hamiltonian H_2 is

$$H_2' = \sum_{j,\ell,p}\sum_{m,n} \left[\sigma a_j^+ (a_\ell)^m (a_p)^n + \sigma^* a_j (a_\ell^+)^m (a_p^+)^n\right] . \tag{4.47}$$

The corresponding operator of the rate of change of the occupation number of the i-th mode (in the interaction representation) is

$$(\dot{n}_i')_{\mathrm{I}} = -\frac{\mathrm{i}}{\hbar} \exp\left(\frac{\mathrm{i}}{\hbar} H_0 t\right) [n_i,\, H_2'] \exp\left(-\frac{\mathrm{i}}{\hbar} H_0 t\right) . \tag{4.48}$$

The commutation relation $[n_i,\, H_2']$ in Eq. (4.48) is evaluated using relations of the type

$$\begin{aligned}\left[a_i^+ a_i,\, a_j^+ (a_\ell)^m (a_p)^n\right] &= -m\delta_{i\ell} a_j^+ (a_\ell)^m (a_p)^n - n\delta_{ip} a_j^+ (a_\ell)^m (a_p)^n \\ &\quad + \delta_{ij} a_j^+ (a_\ell)^m (a_p)^n \end{aligned} \tag{4.49}$$

where Eq. (13.19) (Appendix 13.2) was used. Making use of Eqs (4.49) and (13.18) we get from Eq. (4.48)

$$\begin{aligned}(\dot{n}_i')_{\mathrm{I}} = -\frac{\mathrm{i}}{\hbar} \sum_{j,\ell,p}\sum_{m,n} \Big(&\sigma \left\{-m\delta_{i\ell} a_j^+ (a_\ell)^m (a_p)^n - n\delta_{ip} a_j^+ (a_\ell)^m (a_p)^n \right. \\ &\left. + \delta_{ij} a_j^+ (a_\ell)^m (a_p)^n\right\} \exp\left[\mathrm{i}(\omega_j - m\omega_\ell - n\omega_p)t\right] \\ &+ \sigma^* \left\{m\delta_{i\ell} a_j (a_\ell^+)^m (a_p^+)^n + n\delta_{ip} a_j (a_\ell^+)^m (a_p^+)^n \right. \\ &\left. - \delta_{ij} a_j (a_\ell^+)^m (a_p^+)^n\right\} \exp\left[\mathrm{i}(-\omega_j + m\omega_\ell + n\omega_p)t\right]\Big) . \end{aligned} \tag{4.50}$$

Using the adiabatic hypothesis and the operator of the dynamic evolution we determine the perturbed state vector. From Eq. (4.13) where the Hamiltonian H_1 is replaced by H_2' we determine the interaction Hamiltonian

$$(H_2')_{\mathrm{I}} = \exp(st) \sum_{j,\ell,p} \sum_{m,n} \{\sigma a_j^+ (a_\ell)^m (a_p)^n \exp[\mathrm{i}(\omega_j - m\omega_\ell - n\omega_p)t] + \sigma^* a_j (a_\ell^+)^m (a_p^+)^n \exp[\mathrm{i}(-\omega_j + m\omega_\ell + n\omega_p)t]\} . \quad (4.51)$$

Operator of the dynamic evolution is given by Eq. (13.26); this operator in the first order approximation has the form

$$\begin{aligned} U(t,-\infty) &= 1 - \frac{\mathrm{i}}{\hbar}\int_{-\infty}^{t} (H_2')_{\mathrm{I}} \mathrm{d}t_1 \\ &= 1 - \frac{1}{\hbar}\sum_{j,\ell,p}\sum_{m,n}\left\{\sigma a_j^+ (a_\ell)^m (a_p)^n \frac{\exp[\mathrm{i}(\omega_j - m\omega_\ell - n\omega_p)t]}{(\omega_j - m\omega_\ell - n\omega_p) - \mathrm{i}s}\right. \\ &\left. + \sigma^* a_j (a_\ell^+)^m (a_p^+)^n \frac{\exp[\mathrm{i}(-\omega_j + m\omega_\ell + n\omega_p)t]}{(-\omega_j + m\omega\ell + n\omega_p) - \mathrm{i}s}\right\} \end{aligned} \quad (4.52)$$

and changing summation over ω_j to integration (similarly as we did in derivation of Eq. (4.16)) we get

$$|\Psi_{\mathrm{I}}(t)\rangle = \left\{1 - \pi\frac{\mathrm{i}}{\hbar}\sum_{\ell,p}\sum_{m,n}\left[\sigma a^+_{\ell m + pn}(a_\ell)^m (a_p)^n + \sigma^* a_{-\ell m - pn}(a_\ell^+)^m (a_p^+)^n\right]\right\} |\Psi_0\rangle . \quad (4.53)$$

The expectation value $\langle(\dot{n}_i')_{\mathrm{I}}\rangle$ is determined from Eq. (4.50), Eq. (4.53) for $|\Psi_{\mathrm{I}}(t)\rangle$, and from a similar relation for $\langle\Psi_{\mathrm{I}}(t)|$. We get

$$\begin{aligned} \langle(\dot{n}_i')_{\mathrm{I}}\rangle &= \langle\Psi_{\mathrm{I}}(t)|(\dot{n}_i')_{\mathrm{I}}|\Psi_{\mathrm{I}}(t)\rangle \\ &= \langle\Psi_0|\left\{1 + \pi\frac{\mathrm{i}}{\hbar}\sum_{q,r}\sum_{m,n}\left[\sigma^* a_{-qm-rn}(a_q^+)^m (a_r^+)^n + \sigma a^+_{qm+rn}(a_q)^m (a_r)^n\right]\right\} \times \\ &\left(-\frac{\mathrm{i}}{\hbar}\right)\sum_{j,\ell,p}\sum_{m,n}\left\{\sigma\left[-m\delta_{i\ell}a_j^+(a_\ell)^m(a_p)^n - n\delta_{ip}a_j^+(a_\ell)^m(a_p)^n\right.\right. \\ &\left. + \delta_{ij}a_j^+(a_\ell)^m(a_p)^n\right]\exp[\mathrm{i}(\omega_j - m\omega_\ell - n\omega_p)t] \end{aligned}$$

$$+\sigma^*\left[m\delta_{i\ell}a_j(a_\ell^+)^m(a_p^+)^n + n\delta_{ip}a_j(a_\ell^+)^m(a_p^+)^n\right.$$

$$\left.\left.-\delta_{ij}a_j(a_\ell^+)^m(a_p^+)^n\right]\exp\left[\mathrm{i}(-\omega_j+m\omega_\ell+n\omega_p)t\right]\right\}\times$$

$$\left\{1-\pi\frac{\mathrm{i}}{\hbar}\sum_{q',r'}\sum_{m,n}\left[\sigma a^+_{q'm+r'n}(a_{q'})^m(a_{r'})^n\right.\right.$$

$$\left.\left.+\sigma^* a_{-q'm-r'n}(a_{q'}^+)^m(a_{r'}^+)^n\right]\right\}|\Psi_0\rangle\,. \tag{4.54}$$

We assume that only two fundamental frequencies ω_ℓ and ω_p exist in the system of oscillating units and that the frequency ω_j satisfies the relation

$$\omega_j = m\omega_\ell + n\omega_p \tag{4.55}$$

where $m = 0, 1, 2, 3, \ldots$, $n = 0, 1, 2, 3, \ldots$. We will determine the expectation values $\langle\dot{n}_j\rangle$, $\langle\dot{n}_\ell\rangle$, and $\langle\dot{n}_p\rangle$ corresponding to the perturbation Hamiltonian H_2'. We get from Eq. (4.54)

$$\langle\dot{n}_j\rangle = B\,,\quad \langle\dot{n}_\ell\rangle = -\sum_{n=0}^{+\infty}\left(\sum_{m=0}^{+\infty}\right)' mB\,,\quad \langle\dot{n}_p\rangle = -\sum_{m=0}^{+\infty}\left(\sum_{n=0}^{+\infty}\right)' nB \tag{4.56}$$

where

$$B = \left(\frac{\pi}{\hbar^2}\right)|\sigma|^2\left[(\langle n_j\rangle+1)\,\alpha_\ell(m)\alpha_p(n) - \langle n_j\rangle\beta_\ell(m)\beta_p(n)\right]\,. \tag{4.57}$$

The primes attached to the symbols of summation in Eq. (4.56) denote that the term for $m = 1$ (or for $n = 1$) is omitted if $n = 0$ (or $m = 0$) (the same holds for the frequencies $-m\omega_\ell + n\omega_p$ and $m\omega_\ell - n\omega_p$ given below). The coefficients $\alpha_\ell(m)$ and $\beta_p(n)$ in Eq. (4.57) are given by

$$\begin{aligned}\alpha_\ell(m) &= \langle n_\ell\rangle\,(\langle n_\ell\rangle - 1)\ldots(\langle n_\ell - m + 1\rangle)\\ \beta_p(n) &= (\langle n_p\rangle + n)\,(\langle n_p\rangle + n - 1)\ldots(\langle n_p\rangle + 1)\,.\end{aligned} \tag{4.58}$$

We have to emphasize that the frequency of the vibration mode j is given by Eq. (4.55). The index j in Eqs (4.55)–(4.57) is not arbitrary and may be written as $j = \ell m + pn$.

We evaluated only the first two terms of the perturbation Hamiltonian H_2 in Eq. (4.45). Using the same procedure as the above we may determine the expectation values of the terms ϱ and ξ in Eq. (4.45).

The ϱ term has the combination frequency

$$\omega_j = -m\omega_\ell + n\omega_p \tag{4.59}$$

(where $j = -\ell m + pn$) and the expectation values $\langle \dot{n}_j \rangle$, $\langle \dot{n}_\ell \rangle$, and $\langle \dot{n}_p \rangle$ are given by

$$\langle \dot{n}_j \rangle = C\,, \quad \langle \dot{n}_\ell \rangle = \sum_{n=0}^{+\infty} \sum_{m=0}^{m_0} mC\,, \quad \langle \dot{n}_p \rangle = -\sum_{m=0}^{+\infty} \left(\sum_{n=n_0}^{+\infty} \right)' nC \tag{4.60}$$

where $0 \leq m_0 < n\omega_p/\omega_\ell$, $n_0 > m\omega_\ell/\omega_p$ (as follows from Eq. (4.59) for $\omega_j > 0$), and

$$C = \left(\frac{\pi}{\hbar^2} \right) |\varrho|^2 \left[(\langle n_j \rangle + 1)\, \beta_\ell(m) \alpha_p(n) - \langle n_j \rangle \alpha_\ell(m) \beta_p(n) \right] . \tag{4.61}$$

The ξ term in Eq. (4.45) has the combination frequency

$$\omega_j = m\omega_\ell - n\omega_p \tag{4.62}$$

(where $j = \ell m - pn$) and the expectation values $\langle \dot{n}_j \rangle$, $\langle \dot{n}_\ell \rangle$, and $\langle \dot{n}_p \rangle$ are given by

$$\langle \dot{n}_j \rangle = D\,, \quad \langle \dot{n}_\ell \rangle = -\sum_{n=0}^{+\infty} \left(\sum_{m=m_0}^{+\infty} \right)' mD\,, \quad \langle \dot{n}_p \rangle = \sum_{m=0}^{+\infty} \sum_{n=0}^{n_0} nD \tag{4.63}$$

where $m_0 > n\omega_p/\omega_\ell$, $0 \leq n_0 < m\omega_\ell/\omega_p$ (as follows from Eq. (4.62) for $\omega_j > 0$), and

$$D = \left(\frac{\pi}{\hbar^2} \right) |\varrho|^2 \left[(\langle n_j \rangle + 1)\, \alpha_\ell(m) \beta_p(n) - \langle n_j \rangle \beta_\ell(m) \alpha_p(n) \right] . \tag{4.64}$$

We have to remark that the expressions for B, C, and D given by Eqs (4.57), (4.61), and (4.64), respectively, may need corrections. Eq. (4.53) was derived under the assumption that the density of states equals 1. (The same assumption was made when we derived Eq. (4.16)). This assumption need not be valid. The density of states may be a function of ω_j. For a well-behaved function of the density of states with no singular points only the value of the residue of the function given by Eq. (4.52) is changed. The final effect is the change of transition probabilities and, therefore, the expressions for B, C, and D have to be multiplied by a coefficient corresponding to the new value of the residue.

The energy transfer between particular modes whose frequencies are in convenient relationships is described by Eqs (4.56), (4.60), and (4.63). We may write a rate equation which includes the terms representing the perturbation Hamiltonians H_1' and H_2 in the form

$$\langle \dot{n}_i \rangle = \langle s_i \rangle + \langle n_i \rangle_\Phi + A_i = \langle s_i \rangle - \Phi_i \left[\langle n_i \rangle \exp\left(\frac{\hbar\omega_i}{kT}\right) - (\langle n_i \rangle + 1) \right] + A_i \tag{4.65}$$

where

$$\begin{aligned} A_i &= \langle \dot{n}_j \rangle \quad &&\text{for} \quad i = j = \pm \ell m \pm pn \\ A_i &= \langle \dot{n}_\ell \rangle \quad &&\text{for} \quad i = \ell \\ A_i &= \langle \dot{n}_p \rangle \quad &&\text{for} \quad i = p\,. \end{aligned} \tag{4.66}$$

A_i in Eqs (4.65) and (4.66) stands for the multiple quanta term which may be determined from Eqs (4.56), (4.60), and (4.63).

The rate equation (4.65) describes energy transfer caused by the non-linear mechanism with no participation of the heat bath. Coupling to the heat bath is given by the Φ term which represents one quantum transfer. Energy transfer between vibration modes depends on the non-linear properties of the vibration system. This mechanism of energy transfer between vibration modes strongly depends on the frequency relations as follows from Eqs (4.55), (4.59), and (4.62). The frequency of anharmonic oscillators depends on excitation, i.e. on the energy level. Therefore, at certain energy level the frequency relations for the energy transfer may be satisfied but a change of the energy level can cut off the energy transfer.

Apart from the multipliers m and n the amount of energy transferred by the non-linear mechanism is determined by Eqs (4.57), (4.61), and (4.64) for B, C, and D, respectively. These quantities depend on the occupation numbers. The right hand side of each of the relations for B, C, and D is given by the difference of two terms. The rate of energy supply depends on the absolute value of the difference and its sign (i.e. if the value is positive or negative) determines the direction of energy transfer between the modes. Therefore, the direction of energy transfer depends on the excitation of the vibration modes. The multiple quanta interaction processes with no participation of the heat bath have the properties similar to the Fröhlich χ process – they may transfer energy downwards to the low frequency modes as well as upwards to the high frequency modes. Both the processes may be important to create a state far from thermodynamic equilibrium.

4.5 Power Relations

The expectation values $\langle \dot{n}_j \rangle$, $\langle \dot{n}_\ell \rangle$, and $\langle \dot{n}_p \rangle$ given by Eqs (4.56), (4.60), and (4.63) multiplied by the frequency and Planck's constant determine the change of the energy in the unit time, i.e. the power which changes the energy levels of vibration modes. Therefore, these power relations make possible to examine the input power and the output power of the vibration modes. First of all we will use Eq. (4.56) which holds for the frequency $\omega_j = m\omega_\ell + n\omega_p$ and the momentum denoted by $j = \ell m + pn$. We multiply the first relation in Eq. (4.56) by m and summ over m and n. The result has the same absolute value as $\langle \dot{n}_\ell \rangle$, the second relation in (4.56). We get

$$\sum_{n=0}^{+\infty} \left(\sum_{m=0}^{+\infty} \right)' m\langle \dot{n}_j \rangle + \langle \dot{n}_\ell \rangle = 0 \tag{4.67}$$

where the prime denotes that $m = 1$ is omitted in summation if $n = 0$. Eq. (4.67) may be written in the form

$$\hbar \sum_{n=0}^{+\infty} \sum_{m=0}^{+\infty} m \frac{\langle \dot{n}_j \rangle (m\omega_\ell + n\omega_p)}{m\omega_\ell + n\omega_p} = \sum_{n=0}^{+\infty} \sum_{m=0}^{+\infty} m \frac{P_{m,n}}{m\omega_\ell + n\omega_p} = 0 \tag{4.68}$$

where

$$P_{m,n} = \hbar \langle \dot{n}_j \rangle (m\omega_\ell + n\omega_p)$$

is the power (positive or negative) connected with the mode $j = \ell m + pn$ (with the frequency $m\omega_\ell + n\omega_p$). A similar relation may be derived after multiplication of the first relation in Eq. (4.56) by n instead of m. The result together with the third relation in Eq. (4.56) yields

$$\hbar \sum_{m=0}^{+\infty} \sum_{n=0}^{+\infty} n \frac{\langle \dot{n}_j \rangle (m\omega_\ell + n\omega_p)}{m\omega_\ell + n\omega_p} = \sum_{m=0}^{+\infty} \sum_{n=0}^{+\infty} n \frac{P_{m,n}}{m\omega_\ell + n\omega_p} = 0\,. \tag{4.69}$$

We may also derive the power relations for the combination frequencies

$$\omega_j = -m\omega_\ell + n\omega_p \; (j = -\ell m + pn), \quad \omega_j = m\omega_\ell - n\omega_p \; (j = \ell m - pn)$$

from Eqs (4.60) and (4.63), respectively. One gets from Eq. (4.60)

$$\hbar \left[\sum_{n=0}^{+\infty} \sum_{m=0}^{m_0} m \frac{\langle \dot{n}_j \rangle (-m\omega_\ell + n\omega_p)}{-m\omega_\ell + n\omega_p} - \langle \dot{n}_\ell \rangle \right] = \sum_{n=0}^{+\infty} \sum_{m} m \frac{P_{-m,n}}{-m\omega_\ell + n\omega_p} = 0 \tag{4.70}$$

and

$$\hbar \sum_{m=0}^{+\infty} \sum_{n=n_0}^{+\infty} n \frac{\langle \dot{n}_j \rangle (-m\omega_\ell + n\omega_p)}{-m\omega_\ell + n\omega_p} = \sum_{m=0}^{+\infty} \sum_{n=n_0}^{+\infty} n \frac{P_{-m,n}}{-m\omega_\ell + n\omega_p} = 0 \tag{4.71}$$

where $n\omega_p \geq m\omega_\ell$, the sum on the right hand side of Eq. (4.70) contains the term for $m = 1$, $n = 0$, and

$$P_{-m,n} = \hbar\langle\dot{n}_j\rangle(-m\omega_\ell + n\omega_p) \ .$$

From Eq. (4.63) one obtains

$$\hbar\sum_{n=0}^{+\infty}\sum_{m=m_0}^{+\infty} m\frac{\langle\dot{n}_j\rangle(m\omega_\ell - n\omega_p)}{m\omega_\ell - n\omega_p} = \sum_{n=0}^{+\infty}\sum_{m=m_0}^{+\infty} m\frac{P_{m,-n}}{m\omega_\ell - n\omega_p} = 0 \tag{4.72}$$

and

$$\hbar\left[\sum_{m=0}^{+\infty}\sum_{n=0}^{n_0} n\frac{\langle\dot{n}_j\rangle(m\omega_\ell - n\omega_p)}{m\omega_\ell - n\omega_p} - \langle\dot{n}_p\rangle\right] = \sum_{m=0}^{+\infty}\sum_{n} n\frac{P_{m,-n}}{m\omega_\ell - n\omega_p} = 0 \tag{4.73}$$

where $n\omega_p \leq m\omega_\ell$, the sum on the right hand side of Eq. (4.73) contains the term for $n = 1$, $m = 0$, and

$$P_{m,-n} = \hbar\langle\dot{n}_j\rangle(m\omega_\ell - n\omega_p) \ .$$

The power relations given by Eqs (4.67)–(4.73) are derived from the expectation values of the rate of change of the occupation numbers of the vibration modes. The power relations correspond to the principle of the conservation of energy. They make possible to assess the spectral energy transformation caused by the non-linear effects in the vibration systems. Similar relations were derived by Manley and Rowe [109] for classical electrical oscillators.

Formally, the power relations for a non-linear system may be derived in the following way. We assume the total energy of the vibration system as well as the energy of the normal modes with different frequencies to be independent of time. Energy losses are negligible or they are compensated by the energy supply. The Hamiltonian of the system

$$H = \sum_i H_i \tag{4.74}$$

(where the index i refers to the mode with the frequency ω_i) is, therefore, independent of time too. The Hamiltonians H_i in Eq. (4.74) include interactions of the i-th mode with the other modes in the system. The time derivative of the expectation value is given by

$$\begin{aligned}\frac{\mathrm{d}}{\mathrm{d}t}\langle\Psi|H|\Psi\rangle &= \frac{\mathrm{d}}{\mathrm{d}t}\langle\Psi|\sum_i H_i|\Psi\rangle = \sum_i\frac{\mathrm{d}}{\mathrm{d}t}\langle\Psi|H_i|\Psi\rangle \\ &= \sum_i\frac{\mathrm{d}}{\mathrm{d}t}E_i = \sum_i P_i = 0\end{aligned} \tag{4.75}$$

where E_i contains the energy component of the mode ω_i. We will use only two fundamental frequencies ω_ℓ and ω_p ($\omega_i = m\omega_\ell + n\omega_p$). We get

$$\frac{\mathrm{d}}{\mathrm{d}t}E_i = P_i = \frac{1}{2\pi}\left(m\omega_\ell + n\omega_p\right)W_{m,n} \tag{4.76}$$

where $W_{m,n}$ denotes the expectation value of the energy stored in the normal mode within one period $T_{m,n} = 2\pi/\left(m\omega_\ell + n\omega_p\right)$. After summation we get from Eqs (4.75) and (4.76)

$$\sum_i P_i = \frac{1}{2\pi}\left(\omega_\ell \sum_m \sum_n mW_{m,n} + \omega_p \sum_m \sum_n nW_{m,n}\right) = 0\,. \tag{4.77}$$

Eq. (4.77) equals zero also in the case that both the terms on the right hand side of this equation equal zero independently. The integer multipliers m and n may be positive, negative, or zero. After substitution of

$$W_{m,n} = P_{m,n}T_{m,n} = 2\pi\frac{P_{m,n}}{m\omega_\ell + n\omega_p}$$

(where $P_{m,n}$ stands for P_i with the angular frequency $m\omega_\ell + n\omega_p$) into Eq. (4.77) we get

$$\sum_m \sum_n m\frac{P_{m,n}}{m\omega_\ell + n\omega_p} = 0 \tag{4.78}$$

$$\sum_m \sum_n n\frac{P_{m,n}}{m\omega_\ell + n\omega_p} = 0\,. \tag{4.79}$$

Summations in Eqs (4.78) and (4.79) should be limited to the finite values of m and n.

We assumed that energy of all the normal modes is constant. The energy changes caused by the non-linear mechanism given by Eqs (4.78) and (4.79) have to be compensated by the energy supply or by the energy loss to keep the energy level of each mode unchanged. This condition is necessary for creation of the stationary state. If such state exists Eqs (4.78) and (4.79) express the possibility of describing the non-linear system by superposition of harmonic components.

5. Energy Condensation

5.1 Steady State Solution of the Fröhlich Rate Equation

A detailed description of a Fröhlich system is given in Chapter 4. Here we will point out that the Fröhlich systems are capable of polarization vibrations. The oscillating units may be formed by identical polar vibrators interacting with each other which results in a band of normal modes with different frequencies. Long range interactions within the vibration system as well as between the vibration system and the heat bath depend on the electric nature of the vibrations.

Fröhlich [7,8,75,79–81] considered a band of polarization waves consisting of normal modes with the frequencies

$$\nu_1 \leq \nu_2 \leq \ldots \leq \nu_i \leq \ldots \leq \nu_z \quad \text{where} \quad i = 1, 2, 3, \ldots, z \, . \tag{5.1}$$

The band of normal modes is excited by the energy supply from a source. The total energy supply W_s is given by the relation

$$W_s = \sum_{i=1}^{z} s_i h \nu_i \tag{5.2}$$

where s_i is the number of energy quanta supplied to the mode with the frequency ν_i in the unit time (the rate of energy supply). The occupation number n_i determining the number of energy quanta in the i-th normal mode may be greater than the thermodynamic equilibrium value n_{0i}.

The band of the normal modes is assumed to be embedded in the heat bath. The coupling of the band of normal modes to the heat bath has a fundamental role. Energy is transferred between the band of polarization waves and the heat bath. The heat bath is formed by the environment of the vibration system and we assume that it is in thermodynamic equilibrium. The Coulomb interaction of the vibration system with the heat bath is important for energy transfer. The linear energy transfer to the heat bath may be described by the rate equation

$$(\dot{n}_i)_\Phi = \left(\frac{\partial n_i}{\partial t}\right)_\Phi = -\Phi_i \frac{n_i - n_{0i}}{n_{0i}} \tag{5.3}$$

where Φ_i is the transition probability for the linear energy transfer between the vibration mode with the frequency ν_i and the heat bath. Using Planck's relation

$$n_{0i} = \frac{1}{\exp(\beta\nu_i) - 1} \tag{5.4}$$

(where $\beta = h/kT$, k is the Boltzmann constant and T is the temperature) we get from Eq. (5.3)

$$(\dot{n}_i)_\Phi = -\Phi_i \left[\overbrace{n_i \exp(\beta\nu_i)}^{\uparrow} - \underbrace{(n_i + 1)}_{\uparrow} \right] . \tag{5.5}$$

We will specify the terms on the right hand side of Eq. (5.5). We will not take into consideration the total energy of the Fröhlich system and confine our analysis to the energy stored in the vibrations. The minimum value of the vibration energy is $\frac{1}{2}h\nu$. In the following we will neglect this term. The first term on the right hand side of Eq. (5.5) represents a decrease in the energy of the vibration mode with the frequency ν_i on account of the interaction with the heat bath and the second term represents an increase in the energy of the mode caused by this interaction. Therefore, the term $\Phi_i n_i \exp(\beta\nu_i)$ in Eq. (5.5) stands for the flow of the energy out of the mode and the term $\Phi_i(n_i + 1)$ for the flow in the opposite direction – flow of the energy from the heat bath to the vibration mode. This is visualised by the vertical arrows in Eq. (5.5). The first and the second arrow denotes the flow of energy from the mode and into the mode, respectively. The terms $n_i \exp \beta\nu_i$ and $n_i + 1$ may be regarded as a measure of the probability of energy transfer from and into the i-th mode, respectively.

The system with oscillating charges may be coupled to the heat bath by means of the electric field which can lead to the combination of emission and absorption of energy quanta in two vibration modes with different frequencies. This process has analogy in ionic crystals. Electrons in ionic crystals interact strongly with longitudinal phonons through the electric field of the polarization wave. This is a long range interaction. The combined absorption and emission of the energy quantum by an electron may be accomplished in a very short time [82]. The emitted quantum of energy need not be equal to the quantum absorbed (but the conservation condition of the energy and of the momentum has to be preserved). The combination of emission and absorption enable energy transfer between the normal modes with different frequencies in the vibration system. The interaction of this type results in energy transfer between the vibration modes (we may assume that this is a non-linear interaction of a special type). The energy transferred from the mode ν_i to the mode ν_j with participation of the heat bath may depend on the expression

$$(\dot{n}_i)_{\chi_{ij}} = -\chi_{ij} n_i \exp(\beta\nu_i)(n_j + 1) \tag{5.6}$$

where χ_{ij} is the transition probability of energy transfer between two modes (from the i-th mode to the j-th mode) with participation of the heat bath. This mechanism of energy transfer has a fundamental role in the Fröhlich theory.

The energy flow in the opposite direction depends similarly on the expression

$$(\dot{n}_i)_{\chi_{ji}} = \chi_{ji} n_j \exp(\beta\nu_j)(n_i + 1) \tag{5.7}$$

where χ_{ji} is the transition probability of energy transfer from the j-th mode to the i-th mode with participation of the heat bath. In thermodynamic equilibrium energy transfer from the i-th mode to the j-th mode has to be equal to energy transfer in the opposite direction (the detailed balance condition is fulfilled), i.e.

$$(\dot{n}_i)_{\chi_{ij}} = (\dot{n}_i)_{\chi_{ji}}$$

and from Eqs (5.6) and (5.7) we get

$$\chi_{ij} n_{0i} \exp(\beta\nu_i)(n_{0j} + 1) = \chi_{ji}(n_{0i} + 1) n_{0j} \exp(\beta\nu_j) . \tag{5.8}$$

After replacement of n_{0i} and n_{0j} by the expression given in Eq. (5.4) we get that

$$\chi_{ij} = \chi_{ji} . \tag{5.9}$$

The non-linear coupling of the vibration system to the heat bath results in energy transfer between vibration modes with different frequencies. As follows from Eqs (5.6), (5.7), and (5.9) the term describing energy transfer between normal modes with different frequencies is given by

$$(\dot{n}_i)_\chi = -\sum_j \chi_{ij}\left[n_i(n_j + 1)\exp(\beta\nu_i) - (n_i + 1)n_j \exp\beta\nu_j\right] . \tag{5.10}$$

Using Eqs (5.5) and (5.10) we may write the Fröhlich rate equation in the form

$$\begin{aligned} \dot{n}_i &= s_i - \Phi_i\left[n_i \exp(\beta\nu_i) - (n_i + 1)\right] \\ &\quad -\sum_j \chi_{ij}\left[n_i(n_j + 1)\exp(\beta\nu_i) - (n_i + 1)n_j \exp(\beta\nu_j)\right] \end{aligned} \tag{5.11}$$

where $s_i = S_i\nu_1/\nu_i$ (S_i is the coefficient of energy supply). If no energy is supplied (i.e. $s_i = 0$) the system is in thermodynamic equilibrium. In this case the third term on the right hand side of Eq. (5.11) (the χ term) equals zero as well as the second term (the Φ term) and, therefore, $\dot{n}_i = 0$. We will examine the system for $s_i \neq 0$.

We try to find the stationary solution ($\dot{n}_i = 0$) of the rate equation (5.11). We will write the occupation number n_i in the form

$$n_i = n_{0i} + \Delta n_i \tag{5.12}$$

where Δn_i is that part of the occupation number exceeding the thermodynamic equilibrium value n_{0i}.

Using Eqs (5.11) and (5.12) the total rate of energy supply given by Eq. (5.2) may be written in the form

$$\begin{aligned} W_s &= \sum_i s_i h\nu_i = \sum_i \Phi_i \left[n_i \exp(\beta\nu_i) - (n_i+1)\right] h\nu_i \\ &= \sum_i \Phi_i \left\{n_{0i}\left[\exp(\beta\nu_i) - 1\right] - 1 + \Delta n_i \left[\exp(\beta\nu_i) - 1\right]\right\} h\nu_i \\ &= \sum_i \Phi_i \Delta n_i \left[\exp(\beta\nu_i) - 1\right] h\nu_i \le \Phi_{\max}\left[\exp(\beta\nu_z) - 1\right] h \sum_i \Delta n_i \nu_i \end{aligned} \tag{5.13}$$

where $\Phi_{\max}$ is the maximum value of the Φ_i's. The W_s value depends on the energy supply and represents the excitation condition of the vibration system. The W_s value determines the shift of the vibration system from thermodynamic equilibrium.

Under the stationary conditions ($\dot{n}_i = 0$) we get from Eq. (5.11) the relation

$$\begin{aligned} n_i &= \frac{s_i + \Phi_i + \sum_j \chi_{ij} n_j \exp(\beta\nu_j)}{\left[\Phi_i + \sum_j \chi_{ij}(n_j+1)\right] \exp(\beta\nu_i) - \left[\Phi_i + \sum_j \chi_{ij} n_j \exp(\beta\nu_j)\right]} \\ &= \left[1 + \frac{s_i}{\Phi_i + \sum_j \chi_{ij} n_j \exp(\beta\nu_j)}\right] \frac{1}{\exp\left[\beta(\nu_i - \mu_i)\right] - 1} \end{aligned} \tag{5.14}$$

where the excitation potential μ_i (a quantity similar to the chemical or the electrochemical potential) is given by

$$\begin{aligned} \exp(-\beta\mu_i) &= \frac{\Phi_i + \sum_j \chi_{ij}(n_j+1)}{\Phi_i + \sum_j \chi_{ij} n_j \exp(\beta\nu_j)} \\ &= 1 - \frac{\sum_j \chi_{ij}\left\{n_j\left[\exp(\beta\nu_j) - 1\right] - 1\right\}}{\Phi_i + \sum_j \chi_{ij} n_j \exp(\beta\nu_j)} \ . \end{aligned} \tag{5.15}$$

Using Eqs (5.4) and (5.12) one may write

$$n_j\left[\exp(\beta\nu_j)-1\right]-1=\frac{n_j}{n_{0j}}-1=\Delta n_j\left[\exp(\beta\nu_j)-1\right] \tag{5.16}$$

and Eq. (5.15) can be written in the form

$$\exp(-\beta\mu_i)=1-\frac{\sum_j \chi_{ij}\Delta n_j\left[\exp(\beta\nu_j)-1\right]}{\Phi_i+\sum_j \chi_{ij}n_j\exp(\beta\nu_j)}\,. \tag{5.17}$$

As $0\le\Delta n_j<n_j$, $\exp(\beta\nu_j)-1<\exp(\beta\nu_j)$, $\Phi_i>0$, and $\chi_{ij}>0$ we get from Eq. (5.17)

$$1-\exp(-\beta\mu_i)=0 \quad \text{for} \quad \Delta n_j=0$$

and, therefore,

$$\exp(-\beta\mu_i)=1, \quad \mu_i=0$$
$$0<1-\exp(-\beta\mu_i)<1 \quad \text{for} \quad \Delta n_j>0$$

and, therefore,

$$1>\exp(-\beta\mu_i)>0, \quad \mu_i>0\,. \tag{5.18}$$

In thermodynamic equilibrium $\Delta n_i=0$ and the excitation potential $\mu_i=0$. The occupation number n_i is always positive and cannot increase without limitation. The excitation potential μ_i has to satisfy the relation

$$0\le\mu_i\le\nu_i\,. \tag{5.19}$$

The excitation potential μ_i is always smaller than the frequency ν_i of the normal mode in question.

Using Eq. (5.17) we get from Eq. (5.14)

$$n_i=\left\{1+s_i\frac{1-\exp(-\beta\mu_i)}{\sum_j \chi_{ij}\Delta n_j\left[\exp(\beta\nu_j)-1\right]}\right\}\frac{1}{\exp\left[\beta(\nu_i-\mu_i)\right]-1}\,. \tag{5.20}$$

We assume that $\Phi_i=\Phi$ and $\chi_{ij}=\chi$ for all normal modes of the vibration system. Then

$$\mu_i=\mu, \qquad 0\le\mu\le\nu_1 \tag{5.21}$$

as follows from Eqs (5.17) and (5.19), respectively.

From Eq. (5.13) one can get

$$Z_s=\sum_i s_i=\sum_i \Phi\Delta n_i\left[\exp(\beta\nu_i)-1\right] \tag{5.22}$$

and for $s_i = s$ Eq. (5.20) has the form

$$
\begin{aligned}
n_i &= \left\{1 + \frac{s\Phi}{Z_s\chi}\left[1 - \exp\left(-\beta\mu\right)\right]\right\} \frac{1}{\exp\left[\beta(\nu_i - \mu)\right] - 1} \\
&= A \frac{1}{\exp\left[\beta(\nu_i - \mu)\right] - 1}
\end{aligned}
\tag{5.23}
$$

where

$$
A = 1 + \frac{s\Phi}{Z_s\chi}\left[1 - \exp\left(-\beta\mu\right)\right]
$$

is the amplitude.

If the energy supply to the vibration system is high enough the excitation potential μ can closely approach ν_1. In this case the n_1 value becomes very large as follows from Eq. (5.23). Energy is condensed in the lowest frequency normal mode. The vibration system is far from thermodynamic equilibrium. Energy condensation depends on the term χ in Eq. (5.11). This term represents non-linear coupling to the heat bath.

Energy transfer of the χ type caused by the combination of emission and absorption of energy quanta (non-linear coupling to the heat bath) is a fundamental property of the Fröhlich vibration systems and a necessary condition for energy condensation.

5.2 Extinction of Energy Condensation

The χ term in the Fröhlich rate equation (5.11) contains two quanta processes combining emission and absorption between two normal modes with participation of the heat bath. The processes combining emission–emission and absorption–absorption between two normal modes of the vibration system with different frequencies (with participation of the heat bath) are not included in the Fröhlich rate equation (5.11). These processes can be represented by

$$
(\dot{n}_i)_\Xi = -\sum_j \Xi_{ij}\left\{n_i n_j \exp\left[\beta(\nu_i + \nu_j)\right] - (n_i + 1)(n_j + 1)\right\} \tag{5.24}
$$

where Ξ is the transition probability. Eq. (5.24) is based on a rationale similar to that behind Eq. (5.10). Taking into consideration these processes of energy transfer the total energy transfer between normal modes is given by

$$
\begin{aligned}
(\dot{n}_i)_\chi + (\dot{n}_i)_\Xi = &-\sum_j \chi_{ij}\left[n_i(n_j + 1)\exp\left(\beta\nu_i\right) - (n_i + 1)n_j \exp\left(\beta\nu_j\right)\right] \\
&-\sum_j \Xi_{ij}\left\{n_i n_j \exp\left[\beta(\nu_i + \nu_j)\right] - (n_i + 1)(n_j + 1)\right\}
\end{aligned}
\tag{5.25}
$$

where $(\dot{n}_i)_\chi$ is given by Eq. (5.10). The transition probabilities χ_{ij} and Ξ_{ij} in Eq. (5.25) represent certain average values depending on many elementary microprocesses of energy transfer. The elementary microprocesses underlying χ_{ij} and Ξ_{ij} are different (as well as their probabilities) and, therefore, in a general case $\chi_{ij} \neq \Xi_{ij}$. Nevertheless, in some extraordinary cases this condition need not be satisfied and $\chi_{ij} = \Xi_{ij}$. In this case one gets from Eq. (5.25) that

$$\begin{aligned}(\dot{n}_i)_\chi + (\dot{n}_i)_\Xi &= -\left[n_i \exp(\beta\nu_i) - (n_i+1)\right] \sum_j \chi_{ij} \left[n_j \exp(\beta\nu_j) + (n_j+1)\right] \\ &= -A_i \left[n_i \exp(\beta\nu_i) - (n_i+1)\right] \end{aligned} \tag{5.26}$$

where

$$A_i = \sum_j \chi_{ij} \left[n_j \exp(\beta\nu_i) - (n_j+1)\right] . \tag{5.27}$$

Eq. (5.11) with the term given by Eq. (5.26) has the form

$$\dot{n}_i = s_i - \Phi_i \left[n_i \exp(\beta\nu_i) - (n_i+1)\right] - A_i \left[n_i \exp(\beta\nu_i) - (n_i+1)\right] . \tag{5.28}$$

Eq. (5.28) yields the stationary solution ($\dot{n}_i = 0$) in the form

$$n_i = \frac{s_i + \Phi_i + A_i}{\Phi_i + A_i} \frac{1}{\exp(\beta\nu_i) - 1} . \tag{5.29}$$

The stationary solution given by Eq. (5.29) does not contain the excitation potential. The energy supply to the vibration system affects only the amplitude term A_i. Solution given by Eq. (5.29) differs fundamentally from the solution given by Eq. (5.23); there is no energy condensation in the lowest frequency mode.

Livshits tacitly assumed that $\chi_{ij} = \Xi_{ij}$ and from Eq. (5.29) drew a conclusion that energy condensation in a vibration system cannot exist [105,106]. Fröhlich [80] and Wu and Austin [86] pointed out that these transition probabilities are not equal. Yushina [110] proved that $\chi_{ij} = \Xi_{ij}$ using an assumption that certain macroscopic coefficients determining the χ_{ij} and the Ξ_{ij} are equal. But these macroscopic coefficients contain the coupling constants κ and ξ as follows from Eqs (4.39) and (4.42). κ and ξ represent probability of the microprocesses of absorption–emission (the χ term) and probability of the microprocesses of absorption–absorption and of emission–emission (the Ξ term). Interaction of the polar modes strongly favours absorption–emission processes and, therefore, the probability of these microprocesses are not equal as is discussed in [75,91,111].

5.3 Computer Simulation of Energy Condensation

The behaviour of a vibration system and its main properties may be derived from the Fröhlich rate equation. The qualitative analysis of the Fröhlich rate equation reveals energy condensation in the lowest frequency mode for a sufficiently high energy supply. To get quantitative results and to evaluate the effect of various parameters and conditions numerical analysis of the Fröhlich rate equation has to be carried out [112–114].

We will use Eq. (5.11) which yields z equations for z normal modes. The computation method in [114] is based on the expansion of n_i into a power series. The first differential coefficient $\dot{n}_i = n_i^{(1)}$ given by Eq. (5.11) depends on the values n_i, n_j, and not on their derivatives. The second differential coefficients $n_i^{(2)}$ are given by the relations of the type

$$\begin{aligned} n_i^{(2)} &= s_i^{(1)} - \Phi_i \left[n_i^{(1)} \exp(\beta \nu_i) - n_i^{(1)} \right] \\ &\quad - \sum_j \chi_{ij} \left\{ \left[n_i^{(1)}(n_j + 1) + n_i n_j^{(1)} \right] \exp(\beta \nu_i) \right. \\ &\quad \left. - \left[(n_i + 1) n_j^{(1)} + n_i^{(1)} n_j \right] \exp(\beta \nu_j) \right\} \end{aligned} \tag{5.30}$$

which depend on the n_i and the n_j values and their first derivatives. The p-th derivatives depend on the derivatives of the highest order $p - 1$. The increments of the energy in a particular mode may be obtained from

$$\Delta n_i(t_k) = \sum_{p=1}^{\infty} \frac{1}{p!} [n_i(t_k)]^{(p)} (\Delta t_k)^p \tag{5.31}$$

where t_k is the time. The occupation number at the time $t_k + \Delta_k$ is given by

$$n_i(t_k + \Delta t_k) = n_i(t_k) + \Delta n_i(t_k) \,. \tag{5.32}$$

Eqs (5.31) and (5.32) may be used to compute the time dependence of the energy condensed in the system provided the convergence condition of Eq. (5.31) is satisfied. We assume that all s_i and n_i quantities are continuous. It follows from the Fröhlich rate equation (5.11), and from equations for the higher order derivatives similar to Eq. (5.30) that these derivatives are continuous too. We further assume that the terms in Eq. (5.31) satisfy the condition

$$\left| [n_i(t_k)]^{(p)} (\Delta t_k)^p \right| < n_i(t_k) \,. \tag{5.33}$$

We can derive relation

$$|n_i^{(1)}| < |s_i| + \Phi_i \left[n_i \exp(\beta\nu_i) + (n_i+1)\right] + \sum_j \chi_{ij} \left[n_i(n_j+1)\exp(\beta\nu_i) + (n_i+1)n_j \exp(\beta\nu_j)\right] = N_i^{(1)} \quad (5.34)$$

and for $s_i^{(1)} = 0$

$$|n_i^{(2)}|\Delta t < \Phi_i \left[n_i \exp(\beta\nu_i) + (n_i+1)\right] + 2\sum_j \chi_{ij} \left[n_i(n_j+1)\exp(\beta\nu_i) + (n_i+1)n_j \exp(\beta\nu_j)\right] < 2N_i^{(1)} \quad (5.35)$$

where n_i, n_j or n_i+1, n_j+1 were substituted for $|n_i^{(1)}|\Delta t$, $|n_j^{(1)}|\Delta t$, respectively. The p-th differential coefficient satisfy the relation

$$|n_i^{(p)}|\Delta t^{p-1} < 2^{p-1} N_i^{(1)} \quad (5.36)$$

(derived for $s_i^{(p)} = 0$). Eqs (5.31) and (5.36) yield

$$|\Delta n_i| < N_i^{(1)} \Delta t \sum_{p=0}^{\infty} \frac{2^p}{(p+1)!} \quad (5.37)$$

which proves the condition of the absolute convergence.

The remainder R after the first m ($m > 1$) terms in Eq. (5.37) can be written in the form

$$R = N_i^{(1)} \Delta t \sum_{p=m}^{\infty} \frac{2^p}{(p+1)!} < N^{(1)} \Delta t \frac{2^m}{(m+1)!} \sum_{p=0}^{\infty} \left(\frac{2}{m+2}\right)^p . \quad (5.38)$$

The R value in Eq. (5.38) for $m = 10$ (13) is at least 10^5 (10^7) times smaller than the sum of the first m terms of the series given on the right hand side of Eq. (5.37). Diminishing the time increment enables us to keep the remainder of the series in Eq. (5.31) as small as necessary. The Lagrange formula may be used to determine the greatest admissible increment.

For high values of condensed energy coupling of the system to the heat bath might be considerably greater than in thermodynamic equilibrium. Increased coupling to the heat bath supresses condensation. Therefore, the transition probabilities Φ and χ depend on the condensed energy, i.e. on the occupation numbers. We will assume that only Φ_i depends on the corresponding occupation number. The non-linear dependence of the Φ_i transition probabilities may be assumed in the form of the upper limit of the energy condensed in the normal modes. The energy over the maximum value $n_{\max}$ is strongly transported to the heat bath. We used several values of $n_{\max}$ – usually $3 \times 10^2\, h\nu_i$ and $3.5 \times 10^3\, h\nu_i$. These values may determine the lower and the upper limit of the range of the occupation numbers in the modes with condensed energy.

Theoretical analysis in this section was used for model calculations of the Fröhlich systems. In the preliminary analyses we employed the power series method and computed the sum of the first 10 or 13 terms in Eq. (5.31) using the time increment $\Delta t_k = 5 \times 10^{-3}$. The power series method provides a simple way of checking the convergence process. Nevertheless, the Runge–Kutta method [115] offers a more efficient numerical treatment of the Fröhlich rate equations than the power series method, which is time consuming. In this case the dependence of Φ on n_i is used in the form

$$\Phi = \Phi_0 \exp\left[\left(\frac{n_i}{n_{\max}}\right)^{10}\right] . \tag{5.39}$$

As both the methods were used for model evaluation we had the possibility of comparing the results. Except for the differences caused by the different dependence of $\Phi = \Phi(n_i)$ the methods yield the same results.

5.4 Frequencies of Coherent Vibrations

Fröhlich originally assumed the vibration system to be located in a cellular membrane [8,75,79,81]. Membrane of a cell is very strongly electrically polarized with the intensity of the "static" electric field of about 10^5 V/cm. Excitation of the vibrations in a particular section of the membrane may occur. The particular section of the membrane represents an oscillating electric dipole and/or multipole. Oscillations with displacement perpendicular to the surface of the membrane have the lowest frequency corresponding to the membrane thickness; half of the wavelength equals the membrane thickness. From the velocity of elastic waves (of the order of 10^3 m/s) and from the thickness of the membrane (of about 10 nm) corresponding to $\lambda/2$ the frequency of vibrations was assessed about 50 GHz. Nevertheless, we assume that vibrations of the lipid bilayer of a membrane can only accompany oscillations excited elsewhere.

A great variety of possible sources with different frequencies may exist. A large number of proteins is embedded in the membrane. At least some proteins in the membrane (as well as in other locations in the cell) are also strongly polarized [116–119]. Oscillations may be excited in the proteins by metabolic energy supply. The oscillating electric field of the proteins can excite the oscillations of the membrane around them. Small sections of the membrane among proteins can oscillate with the frequencies of protein vibrations. The frequencies of stretching, twisting, and bending vibrations in a polypeptide chain are in the infrared region. For instance, the Amide type vibrations in protein molecules [120] occur in the band from about $1000\,\mathrm{cm}^{-1}$ to $3000\,\mathrm{cm}^{-1}$, i.e. from about 30 THz to 90 THz. Frequencies of the giant breathing and rocking modes of biological molecules may be below 10^{11} Hz [75,121]. The lowest frequency ($\nu \neq 0$) of vibrations in microtubules may be of the order of 10^6–10^7 Hz (for a microtubule 10 µm long and velocity of the elastic waves in the range of 10^2–$10^3\,\mathrm{ms}^{-1}$).

Grundler et al. [122–126] and Keilmann [127] measured growth rate of yeast cells (Saccharomyces cerevisiae). After irradiation at certain frequencies in the vicinity of 41 GHz the growth rate was greater than the growth rate with no irradiation. This resonance effect was assumed to be connected with the Fröhlich coherent vibration states [124]. The growth rate of yeast cells after microwave irradiation was measured by Gos et al. too [128]. He did not find any difference between the irradiated and control groups.

For the model systems of the Fröhlich coherent vibration states we assume the frequencies of the normal modes to be in the band from about 40 GHz to 10 THz.

5.5 Models of the Fröhlich System

We will analyze models of vibration systems. The model vibration systems correspond to the systems described by Eq. (5.11). If energy is supplied to the vibration system the time dependence of the energy condensed in the lowest frequency mode is similar (but not equal) to the function $a\,[1 - \exp(-bt)]$ (where a and b are parameters) as is shown in Fig. 5.1 [114]. The time for reaching the steady state conditions depends on the energy supply. The time dependent solutions approach the steady state solutions for $S_i = S$ greater than about $0.1\,\mathrm{s}^{-1}$ and smaller than $10\,\mathrm{s}^{-1}$ within a time period lasting from several minutes up to about one hour after switching on the energy supply. After switching off the energy supply the relaxation time is considerably longer (and can reach several hours).

The model vibration system used in Fig. 5.1 consists of ten normal modes in the frequency range from 0.1 THz to 6.72 THz. The χ_{ij} transition probabilities used for calculation are not uniform. The vibration systems may be composed of several bands in different frequency regions. Interactions between the modes in different frequency bands of vibrations may be weaker than between the modes within the bands. We may expect e.g. that the energy transition between normal modes in the infrared region and in the mm wave region is considerably smaller than within the particular bands. Each frequency band has a normal mode with the lowest frequency (a quasi-lowest frequency mode in the high frequency band). The model system shown in Fig. 5.1 consists of two groups of modes in different frequency regions. A weak interaction between these two groups is assumed resulting in low energy transfer between them. We have to point out that χ_a is the transition probability for energy transfer between the modes of the same group and χ_b is the transition probability for energy transfer between the modes of different groups. The transition probability $\chi_a = 10^{-3}\,\mathrm{s}^{-1}$ and $\chi_b = 0.1\,\chi_a$.

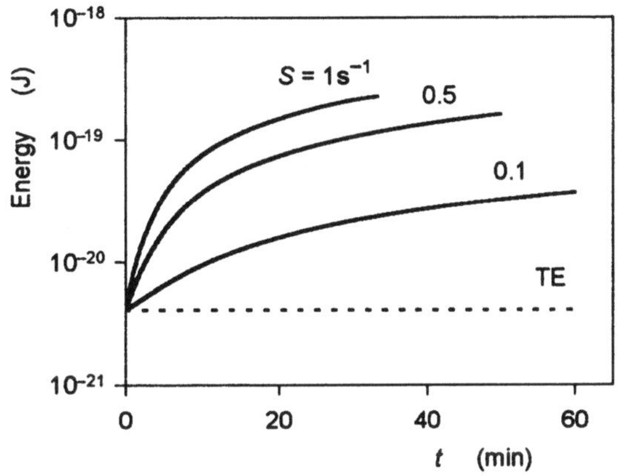

Fig. 5.1 Energy condensed in the lowest frequency mode as a function of the time variable t. Energy supply started at the time $t = 0$. $S_i = S$ is a parameter. $\Phi_i = \Phi = 10^{-2}\,\mathrm{s}^{-1}$, $\chi_a = 10^{-2}\,\mathrm{s}^{-1}$, and $\chi_b = 10^{-3}\,\mathrm{s}^{-1}$. TE – the thermodynamic equilibrium value.

Fig. 5.2 shows computer simulation of a vibration system with ten normal modes in the frequency range from 0.1 THz to 10 THz. Similar models are analyzed in [129]. The vibration system is excited with energy supply ($S_i = S = 5\,\mathrm{s}^{-1}$). The $\chi_{ij} = \chi$ and $\Phi_i = \Phi$ transition probabilities in the whole system are $0.001\,\mathrm{s}^{-1}$ and $0.01\,\mathrm{s}^{-1}$, respectively. We used two values of $n_{\max}$ – 3 500 (a) and 300 (b) – which may determine possible boundaries of condensed energy. Energy is condensed in the lowest frequency mode. Nevertheless, energy in the other modes is higher than in thermodynamic equilibrium. In the model with $n_{\max} = 300$ (Fig. 5.2b) the energy levels of the modes ν_1–ν_4 are nearly the same.

As follows from Eq. (5.23) energy condensation depends on the difference $\nu_1 - \mu$. If the frequency ν_2 of the vibration mode 2 is in the vicinity of the frequency ν_1 the difference $\nu_2 - \mu$ is comparable with the difference $\nu_1 - \mu$ and energy may condense even in the ν_2 mode. Fig. 5.3 shows the case for $\nu_1 = 0.1\,\mathrm{THz}$ and $\nu_2 = 0.11\,\mathrm{THz}$ ($S_i = S = 5\,\mathrm{s}^{-1}$). Energy is always condensed in the lowest frequency mode but also in the mode in its vicinity.

We will analyze models of vibration systems consisting of two groups of modes – the low frequency group of modes and the high frequency group of modes. The χ_a transition probability is $10^{-3}\,\mathrm{s}^{-1}$ for energy transfer between the normal modes of the lower frequency group and for energy transfer between the remaining normal modes (between the modes of the high frequency group). The χ_b transition probability for energy transfer between the normal modes of different groups is $10^{-4}\,\mathrm{s}^{-1}$. Fig. 5.4 shows a model system with 20 normal modes [130]. The low frequency group consists of 13 normal modes [130]. The frequencies of the normal modes in the frequency range from 41.642 GHz to 41.786 GHz are 16 MHz apart. Energy is condensed in all these modes. In this case no significant energy condensation in the quasi-lowest frequency mode (ν_{14}) occurs.

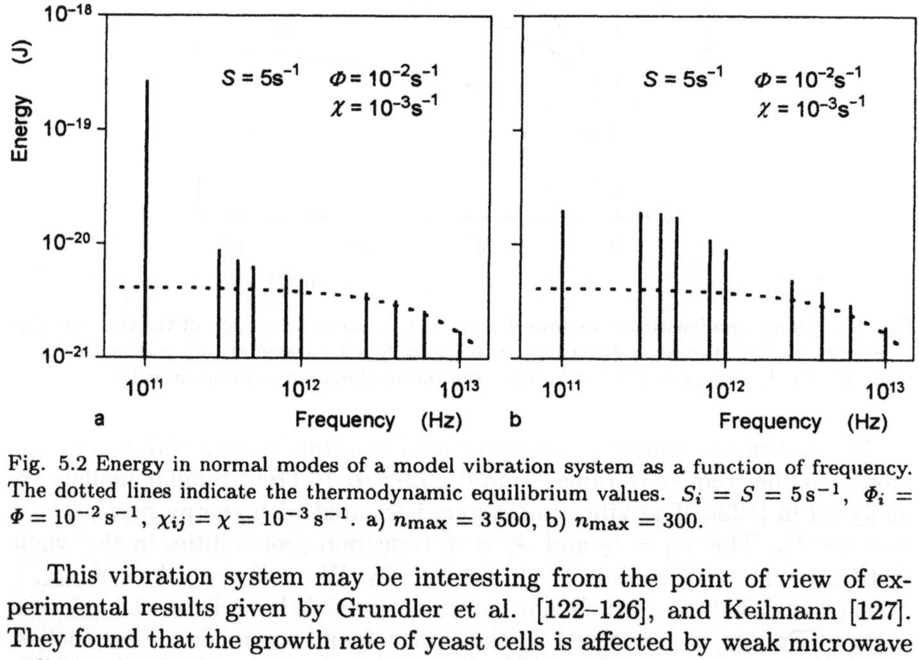

Fig. 5.2 Energy in normal modes of a model vibration system as a function of frequency. The dotted lines indicate the thermodynamic equilibrium values. $S_i = S = 5\,\mathrm{s}^{-1}$, $\Phi_i = \Phi = 10^{-2}\,\mathrm{s}^{-1}$, $\chi_{ij} = \chi = 10^{-3}\,\mathrm{s}^{-1}$. a) $n_{\mathrm{max}} = 3\,500$, b) $n_{\mathrm{max}} = 300$.

This vibration system may be interesting from the point of view of experimental results given by Grundler et al. [122–126], and Keilmann [127]. They found that the growth rate of yeast cells is affected by weak microwave radiation in the frequency range from 41 GHz to 42 GHz. The increased rate of growth and the decreased rate of growth are frequency dependent. The frequencies at which the growth rate is high are 16 MHz apart. As we already noted above Grundler and his collaborators arrived at the conclusion that these findings might support the concept of coherent molecular oscillations and the Fröhlich hypothesis of coherence.

We may assume that the coherent oscillating units are formed e.g. by certain protein molecules in cellular membranes and create coherent oscillation centres which are distributed in cellular membrane. These interacting oscillation centres build up a coherent vibration system of a cell. (A similar oscillation system may be created by the microtubules or by the microtubules together with the membrane proteins.) In the steady state the distribution of the oscillation centres should correspond to the condition of the minimum of energy of the vibration system and of the generated electromagnetic field. In this case one may hardly expect significant absorption of the electromagnetic waves generated by outer sources. On the other hand the distribution of the oscillation centres corresponding to the condition of the minimum of energy need not be preserved in all phases of the cell cycle (e.g. in the mitotic phase this condition need not be satisfied). In this case the electromagnetic emission and absorption may be strong.

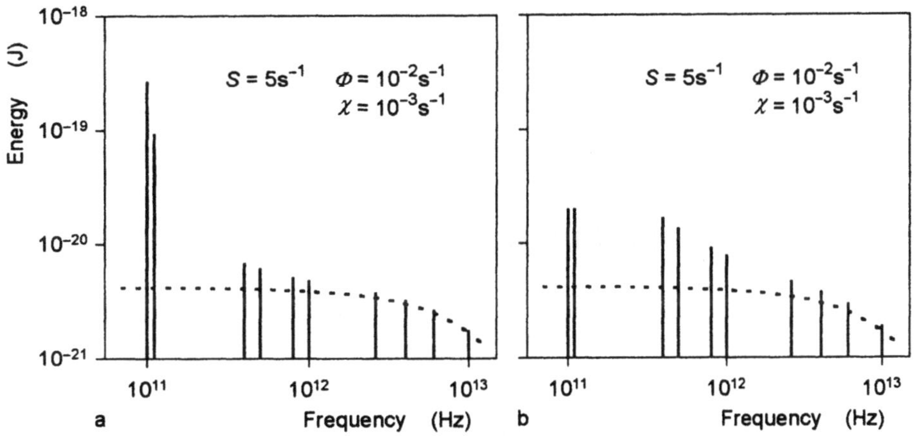

Fig. 5.3 Condensation of energy in the lowest frequency mode and in the normal mode in its close vicinity. The dotted lines indicate the thermodynamic equilibrium values. $S_i = S = 5\,\mathrm{s}^{-1}$, $\Phi_i = \Phi = 10^{-2}\,\mathrm{s}^{-1}$, $\chi_{ij} = \chi = 10^{-3}\,\mathrm{s}^{-1}$. a) $n_{\max} = 3\,500$, b) $n_{\max} = 300$.

Energy transfer between normal modes depends on the non-linear interaction between vibration modes with participation of the heat bath (i.e. on the χ_{ij} transition probabilities). The models of the vibration systems in Figs 5.2 and 5.3 are computed for uniform χ_{ij} transition probabilities in the whole vibration system. In the subsequent models we will use non-uniform χ_{ij} transition probabilities as in Figs 5.1 and 5.4.

Fig. 5.5 shows computer simulation of a model with ten normal modes. The maximum values $n_{\max}$ of the occupation numbers are 3 500 (Figs 5.5a, b, and c) and 300 (Figs 5.5d, e, and f). The low frequency group consists of the first three normal modes (ν_1, ν_2, and ν_3). At the initial time the system was in thermodynamic equilibrium. Energy is always condensed in the lowest frequency mode of the low frequency group of modes (of the lowest frequency mode of the whole system). For a high energy supply ($S_i = S = 10\,s^{-1}$) energy condensation occurs not only in the lowest frequency mode (ν_1) but also in the quasi-lowest frequency mode (ν_4 – the lowest frequency mode of the high frequency group of modes) and in the normal modes in its vicinity (Figs 5.5a and 5.5d). The decreased transfer of energy from a normal mode towards the lower frequency modes of the vibration system is very likely the necessary condition for energy condensation in this mode. For a small energy supply ($S_i = S = 1\,s^{-1}$) energy is condensed in the lowest frequency mode and the energy in the quasi-lowest frequency mode might be near the thermodynamic

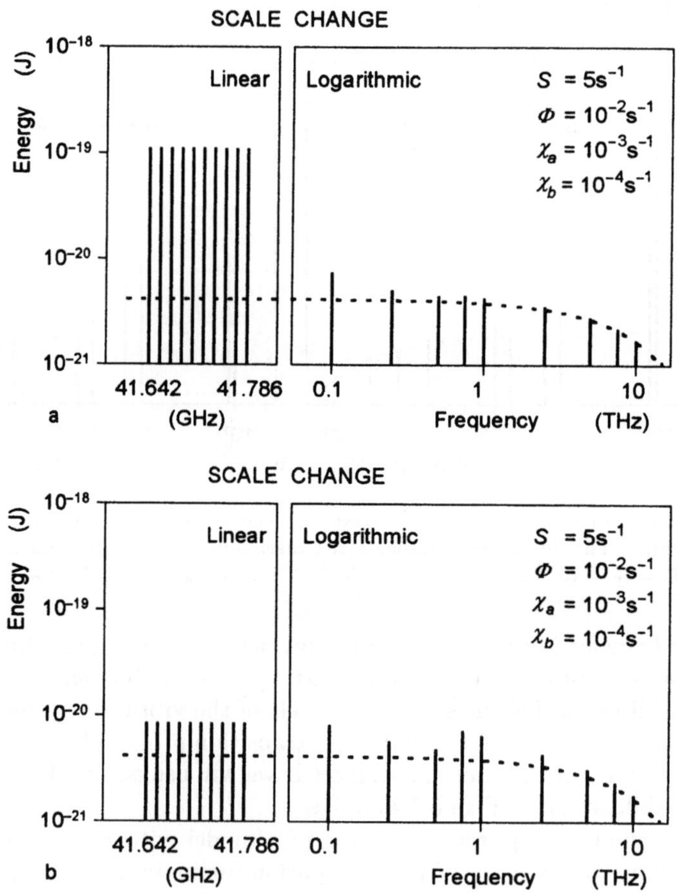

Fig. 5.4 Energy of normal modes as a function of frequency. All the normal modes in the frequency range from 41.642 THz to 41.786 THz are highly excited. $S_i = S = 5\,s^{-1}$, $\Phi_i = \Phi = 10^{-2}\,s^{-1}$, $\chi_a = 10^{-3}\,s^{-1}$, $\chi_b = 10^{-4}\,s^{-1}$. The dotted lines indicate the thermodynamic equilibrium values. a) $n_{max} = 3\,500$, b) $n_{max} = 300$.

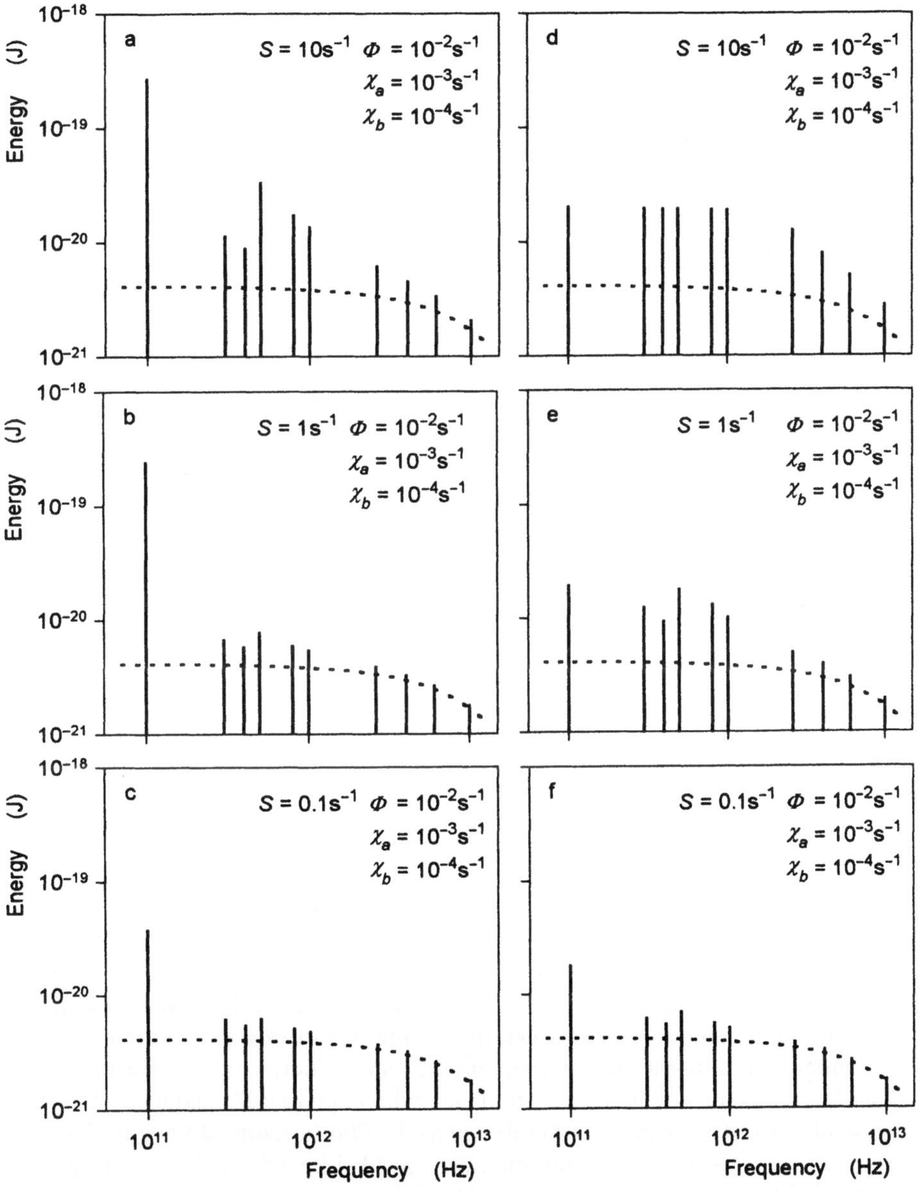

Fig. 5.5 Energy condensed in normal modes as a function of frequency. The dotted lines indicate the thermodynamic equilibrium values.
$\Phi_i = \Phi = 10^{-2}\,s^{-1}$, $\chi_a = 10^{-3}\,s^{-1}$, $\chi_b = 10^{-4}\,s^{-1}$.
$n_{max} = 3\,500$: a) $S_i = S = 10\,s^{-1}$, b) $S_i = S = 1\,s^{-1}$, c) $S_i = S = 0.1\,s^{-1}$.
$n_{max} = 300$: d) $S_i = S = 10\,s^{-1}$, e) $S_i = S = 1\,s^{-1}$, f) $S_i = S = 0.1\,s^{-1}$.

equilibrium value as is shown in Fig. 5.5b for $n_{\max} = 3500$. For $n_{\max} = 300$ energy in this mode does not decrease. For a very small energy supply ($S_i = S = 0.1\,\mathrm{s}^{-1}$) even the energy in the lowest frequency mode decreased (Figs 5.5c and 5.5f).

Fig. 5.6 shows the energy condensed in the lowest frequency mode and in the quasi-lowest frequency mode (in the ν_4 mode) as a function of the coefficient of energy supply S [114]. The model system consits of ten normal modes in the frequency band from 0.1 THz to 6.72 THz; $\chi_a = 10^{-3}\,\mathrm{s}^{-1}$, $\chi_b = 10^{-4}\,\mathrm{s}^{-1}$. The condensed energy increases with increasing S value which is in agreement with the qualitative analysis of the Fröhlich rate equation given by

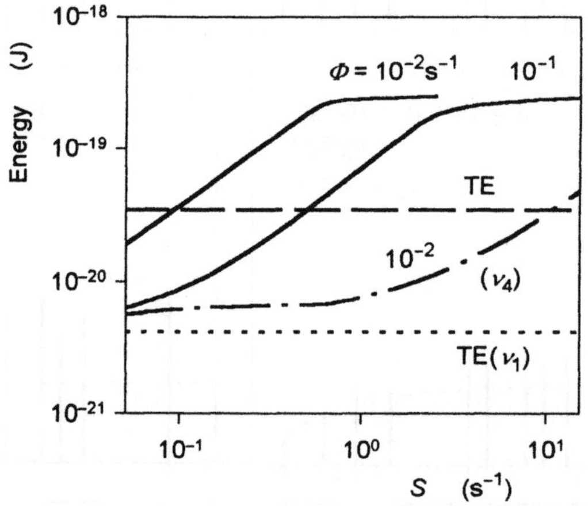

Fig. 5.6 Energy condensed in the lowest frequency mode ν_1 (the full curves) and in the quasi-lowest frequency mode ν_4 (the dashed and dotted curve) as a function of the coefficient of energy supply $S_i = S$. TE and TE(ν_1) denote energy of the whole vibration system and energy of the lowest freqency mode in thermodynamic equilibrium, respectively.

Eqs (5.20) and (5.23). Strong energy condensation occurs provided the energy supply is great enough. The energy condensed in the ν_4 mode is plotted by the dashed and dotted curve in Fig. 5.6. Energy is condensed in the quasi-lowest frequency mode (ν_4) at much higher energy supply than that necessary for condensation in the lowest frequency mode. Energy condensation in the quasi-lowest frequency mode is significant when the lowest frequency mode of the whole vibration system is highly excited. The horizontal lines in Fig. 5.6 represent the energy in thermodynamic equilibrium of the lowest frequency mode – TE(ν_1) and of the whole vibration system – TE.

In a highly excited vibration system energy condensed in some normal modes may be considerably greater than their energy in thermodynamic equilibrium. These modes of the vibration system are far from thermodynamic equilibrium and conditions for coherent vibrations in large regions may be satisfied (giant vibrations). The giant vibrations do not mean big or macroscopic dipole oscillations. Under normal conditions location, space orientation, phase relations, and the other parameters of the oscillators should correspond to the condition of minimum energy.

5.6 Stability of the Steady State Solution

We will analyze stability of the steady state solution of the Fröhlich rate equations. Stability is a property of systems to be resistant to perturbations or disturbances acting upon them. First of all we will outline the criteria of stability (given e.g. in [131]) taking into consideration differential equations describing the Fröhlich system. The system under investigation is in a state of equilibrium when all the coordinates of the solution are constant with respect to time. The Fröhlich system is described by a set of the first order differential equations which do not depend on time explicitly. This autonomous system of differential equations of the type

$$\dot{n}_i = f_i(n_1, n_2, \ldots) \qquad (\text{where} \quad i = 1, 2, \ldots, z) \tag{5.40}$$

has a steady state solution (i.e. solution for $\dot{n}_i = 0$) $\overline{n}_1(t)$, $\overline{n}_2(t), \ldots$. This solution will be called unperturbed. Every other solution is termed perturbed and denoted by $n_1(t)$, $n_2(t), \ldots$. The difference $\eta_i = n_i(t) - \overline{n}_i(t)$ will be referred to as the perturbation. On substituting the perturbations, Eq. (5.40) yields a system of perturbation equations in the form

$$\begin{aligned} \frac{\mathrm{d}\eta_i}{\mathrm{d}t} &= f_i\left[n_1(t), n_2(t), \ldots\right] - f_i\left[\overline{n}_1(t), \overline{n}_2(t), \ldots\right] \\ &= f_i\left[\eta_1 + \overline{n}_1(t), \eta_2 + \overline{n}_2(t), \ldots\right] - f_i\left[\overline{n}_1(t), \overline{n}_2(t), \ldots\right] . \end{aligned} \tag{5.41}$$

The occupation numbers $\overline{n}_1(t)$, $\overline{n}_2(t), \ldots$ are referred to as the coordinates of the steady state point S in a space of states. We assume that ϵ is a small distance from S to the boundary of a domain around this point S. The domain is called the domain of stability. The unperturbed solution $\overline{n}_1(t)$, $\overline{n}_2(t), \ldots$ is stable if for arbitrarily chosen domain of stability ($0 < \epsilon$) we may find a closed surface inside the domain of stability around the point S at a distance $\delta(\epsilon)$ that any trajectory from inside the region enclosed by the surface $\delta(\epsilon)$ never reaches the boundary ϵ. The exact definition of the stability of the periodic solution was given by Lyapunov in the following way [131]: The unperturbed solution $\overline{n}(t)$ is stable if, corresponding to every choice of $0 \leq t_0$ and $0 < \epsilon$, there exists a number $0 < \delta$ such that any perturbation $\eta(t)$ which satisfies $|\eta(t_0)| < \delta$ also satisfies $|\eta(t)| < \epsilon$ for all values of $t_0 \leq t$. The definition of stability leads to the conclusion that the perturbation should satisfy the relation

$$\eta_i = |n_i(t) - \overline{n}_i(t)| \ll \epsilon \tag{5.42}$$

where $0 < \epsilon$ for all i. Eq. 5.42 is the mathematical formulation of the stability condition.

Lyapunov proved that stability of a system described by non-linear differential equations can be assessed by a linear method. Only the linear parts of the set of differential equations are taken into account. We will consider the set of autonomous perturbed equations (5.41) in the form

$$\frac{\mathrm{d}\eta_i}{\mathrm{d}t} = \sum_j \left(\frac{\partial f_i}{\partial n_j}\right)_{\overline{n}_j} (n_j - \overline{n}_j) + \Phi_j(n_1, n_2, \ldots) \tag{5.43}$$

where the first term on the right hand side is linear and the second term

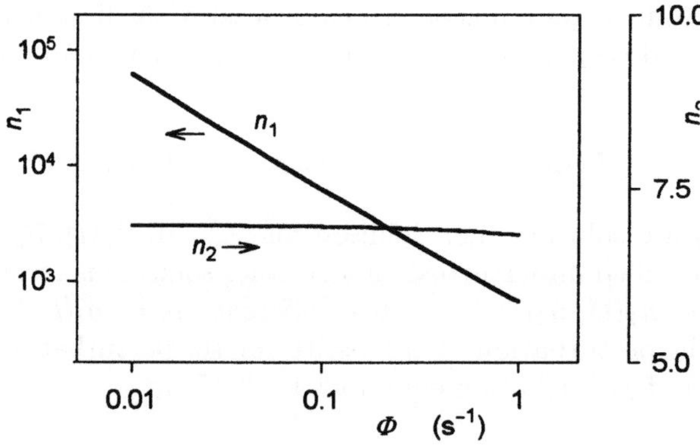

Fig. 5.7 The occupation numbers n_1 and n_2 as a function of the transition probability Φ. $\chi = 0.1\Phi$, $s_1 = s_2 = 5\,\mathrm{s}^{-1}$, $\nu_1 = 0.1\,\mathrm{THz}$, $\nu_2 = 1\,\mathrm{THz}$.

contains all the non-linear terms and can be expanded in a series containing only terms of degree two and higher than two. We denote

$$\left(\frac{\partial f_i}{\partial n_j}\right)_{\overline{n}_j} = a_{ij} \ . \tag{5.44}$$

In a general case $a_{ij} \neq a_{ji}$ and $a_{ij} = 0$ in the case that f_i does not depend on n_j. The linearized system corresponding to Eq. (5.43) is given by

$$\frac{\mathrm{d}\eta_i}{\mathrm{d}t} = \sum_j a_{ij}\eta_j \ . \tag{5.45}$$

Assuming perturbation $\eta_i(t)$ of the occupation number n_i to be in the form

$$\eta_i(t) \approx \exp(\lambda_i t) \tag{5.46}$$

(where λ_i is an unknown constant yet) one gets the characteristic equation of the linearized system (5.45)

$$\lambda_i \eta_i = \sum_j a_{ij} \eta_j \ . \tag{5.47}$$

The characteristic equation (5.47) can be written in the matrix form

$$\begin{bmatrix} a_{11}-\lambda & , & a_{12} & , & \cdots & , & a_{1z} \\ a_{21} & , & a_{22}-\lambda & , & \cdots & , & a_{2z} \\ \vdots & & \vdots & & \vdots & & \vdots \\ a_{z1} & , & a_{z2} & , & \cdots & , & a_{zz}-\lambda \end{bmatrix} \times \begin{bmatrix} \eta_1 \\ \eta_2 \\ \vdots \\ \eta_z \end{bmatrix} = \begin{bmatrix} 0 \\ 0 \\ \vdots \\ 0 \end{bmatrix} . \tag{5.48}$$

The matrix of coefficients a_{ij} on the left hand side of Eq. (5.48) is called

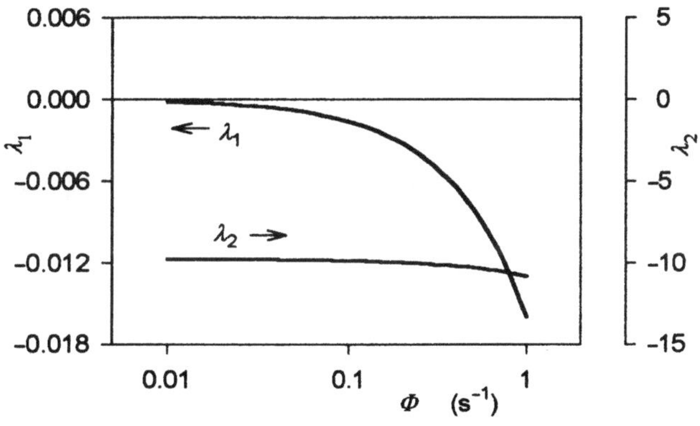

Fig. 5.8 The roots λ_1 and λ_2 of the characteristic equation as a function of the transition probability Φ. The same model as in Fig. 5.7 was used.

the characteristic matrix. A solution of the characteristic equation (5.47) exists if the determinant of the characteristic matrix in Eq. (5.48) equals zero. In this case the roots λ of the characteristic equation can be determined. If all the roots of the characteristic equation of the linearized system of perturbation equations have negative values of the real parts then the unperturbed solution of the complete system is asymptotically stable regardless of the non-linear terms [131]. If some of the roots of the characteristic equation have the real parts positive, the unperturbed solution is unstable regardless of the non-linear terms.

We will analyze a simple system described by two differential equations. The determinant of the characteristic matrix is given by

$$\begin{vmatrix} a_{11}-\lambda & , & a_{12} \\ a_{21} & , & a_{22}-\lambda \end{vmatrix} = 0 \tag{5.49}$$

which yields

$$\lambda^2 - (a_{11} + a_{22})\lambda + (a_{11}a_{22} - a_{12}a_{21}) = 0 \tag{5.50}$$

and

$$\lambda_{1,2} = \frac{a_{11} + a_{22}}{2} \pm \sqrt{\left(\frac{a_{11} + a_{22}}{2}\right)^2 - (a_{11}a_{22} - a_{12}a_{21})}\,. \tag{5.51}$$

Eq. (5.51) will be used to examine the stability of a Fröhlich system with two vibration modes. Under steady state conditions the Fröhlich system is described by the relations

$$\begin{aligned} f_1 = s_1 &- \Phi\left[n_1 \exp\left(\frac{h\nu_1}{kT}\right) - (n_1 + 1)\right] \\ &- \chi\left[n_1(n_2 + 1)\exp\left(\frac{h\nu_1}{kT}\right) - (n_1 + 1)n_2 \exp\left(\frac{h\nu_2}{kT}\right)\right] = 0 \\ f_2 = s_2 &- \Phi\left[n_2 \exp\left(\frac{h\nu_2}{kT}\right) - (n_2 + 1)\right] \\ &- \chi\left[n_2(n_1 + 1)\exp\left(\frac{h\nu_2}{kT}\right) - (n_2 + 1)n_1 \exp\left(\frac{h\nu_1}{kT}\right)\right] = 0 \end{aligned} \tag{5.52}$$

which follow from Eq. (5.11) for $\dot{n}_1 = 0$ and $\dot{n}_2 = 0$. Eq. (5.52) enables us to determine the occupation numbers n_1 and n_2. We get

$$n_1 = \alpha n_2 + \beta\,, \qquad n_2 = -\frac{\gamma}{2} \pm \sqrt{\left(\frac{\gamma}{2}\right)^2 - \kappa} \tag{5.53}$$

where

$$\alpha = -\frac{\exp\left(\frac{h\nu_2}{kT}\right) - 1}{\exp\left(\frac{h\nu_1}{kT}\right) - 1}\,, \qquad \beta = \frac{s_1 + s_2 + 2\Phi}{\Phi\left[\exp\left(\frac{h\nu_1}{kT}\right) - 1\right]}$$

$$\gamma = \frac{\beta}{\alpha} + \frac{\chi \exp\left(\frac{h\nu_2}{kT}\right) - \left\{\chi \exp\left(\frac{h\nu_1}{kT}\right) + \Phi\left[\exp\left(\frac{h\nu_1}{kT}\right) - 1\right]\right\}\alpha}{\chi\left[\exp\left(\frac{h\nu_2}{kT}\right) - \exp\left(\frac{h\nu_1}{kT}\right)\right]\alpha}$$

$$\kappa = \frac{s_1 + \Phi - \left\{\chi \exp\left(\frac{h\nu_1}{kT}\right) + \Phi\left[\exp\left(\frac{h\nu_1}{kT}\right) - 1\right]\right\}\beta}{\chi\left[\exp\left(\frac{h\nu_2}{kT}\right) - \exp\left(\frac{h\nu_1}{kT}\right)\right]\alpha}\,. \tag{5.54}$$

Eqs (5.53) and (5.54) will be used to determine the occupation numbers n_1 and n_2 of a Fröhlich system with two modes whose frequencies are $\nu_1 = 0.1\,\mathrm{THz}$ and $\nu_2 = 1\,\mathrm{THz}$, respectively. The occupation numbers n_1 and n_2 as a function of the transition probability Φ are given in Fig. 5.7. For small values of Φ the occupation number n_1 of the lowest frequency mode is much greater than the occupation number n_2.

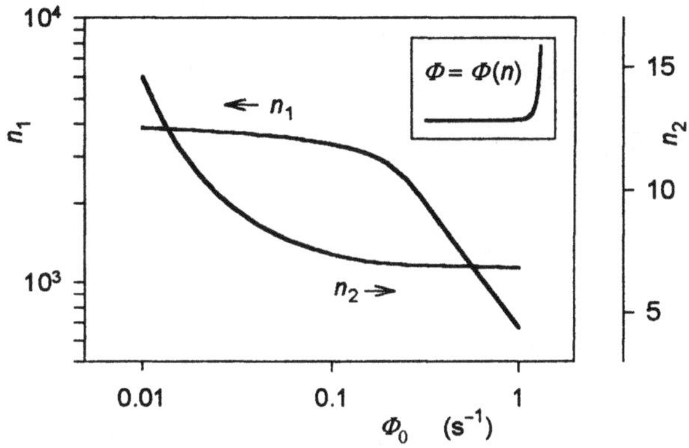

Fig. 5.9 The occupation numbers n_1 and n_2 as a function of Φ_0. The transition probability $\Phi = \Phi(n_i)$ given by Eq. (5.39) is shown in the inset. The other parameters of the model are the same as in Fig. 5.7.

From Eqs (5.52) we determine coefficients of the characteristic matrix

$$
\begin{aligned}
a_{11} &= \frac{\partial f_1}{\partial n_1} = -\Phi\left[\exp\left(\frac{h\nu_1}{kT}\right) - 1\right] \\
&\qquad -\chi\left\{(n_2+1)\exp\left(\frac{h\nu_1}{kT}\right) - n_2\exp\left(\frac{h\nu_2}{kT}\right)\right\} \\
a_{12} &= \frac{\partial f_1}{\partial n_2} = -\chi\left\{n_1\exp\left(\frac{h\nu_1}{kT}\right) - (n_1+1)\exp\left(\frac{h\nu_2}{kT}\right)\right\} \\
a_{21} &= \frac{\partial f_2}{\partial n_1} = -\chi\left\{n_2\exp\left(\frac{h\nu_2}{kT}\right) - (n_2+1)\exp\left(\frac{h\nu_1}{kT}\right)\right\} \\
a_{22} &= \frac{\partial f_2}{\partial n_2} = -\Phi\left[\exp\left(\frac{h\nu_2}{kT}\right) - 1\right] \\
&\qquad -\chi\left\{(n_1+1)\exp\left(\frac{h\nu_2}{kT}\right) - n_1\exp\left(\frac{h\nu_1}{kT}\right)\right\}
\end{aligned}
\tag{5.55}
$$

and from Eq. (5.51) one can evaluate the roots λ_1 and λ_2 of the characteristic equation. Fig. 5.8 shows that both the roots are negative ($\lambda_1 < 0$ and $\lambda_2 < 0$)

for the transition probability Φ in the range $0.01\,\mathrm{s}^{-1} \leq \Phi \leq 1\,\mathrm{s}^{-1}$. The Fröhlich system is asymptotically stable.

The curves in Figs 5.7 and 5.8 were determined under the assumption that Φ and χ are constant and do not depend on excitation. This assumption need not be satisfied. The energy condensed in the lowest frequency mode of the examined Fröhlich model system can reach very high values as follows from Fig. 5.7. For small values of Φ and χ the occupation number n_1 is of the order of 10^4. For high values of condensed energy coupling of the system to the heat bath might be considerably greater than in thermodynamic equilibrium.

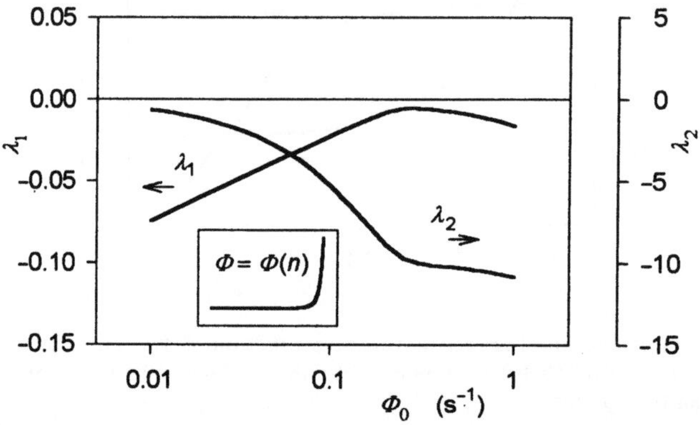

Fig. 5.10 The roots λ_1 and λ_2 of the characteristic equation as a function of Φ_0. The transition probability $\Phi = \Phi(n_i)$ is shown in the inset. The same model as in Fig. 5.9 is used.

Increased coupling to the heat bath suppresses condensation. The transition probabilities Φ and χ are not constant but depend on the condensed energy, i.e. they are functions of the occupation numbers. We will assume that only Φ depends on the corresponding occupation number and use the function given by Eq. (5.39) for $n_{\mathrm{max}} = 3500$. The occupation numbers n_1 and n_2 are shown in Fig. 5.9. Condensation in the lowest frequency mode is strongly limited for $n_{\mathrm{max}} < n_1$ and the energy level remains nearly unchanged. For $\Phi_0 \approx 0.01$ the occupation number n_1 is about one order smaller than in the case given in Fig. 5.7 where Φ does not depend on n_i. The Fröhlich system is asymptotically stable as follows from Fig. 5.10 (λ_1 and λ_2 are negative).

Figs 5.8 and 5.10 show that the Fröhlich systems in question satisfy the requirements of the asymptotic stability under the linearized conditions.

6. Coherence in Systems with Random Energy Supply

6.1 Feedback in the System

In oscillating systems non-linear effects enable energy transformation between spectral lines with different frequencies. Spectral energy transformation is a general property of non-linear systems. Energy supplied to a non-linear vibration system or to a system with non-linear interaction with its surroundings is transferred between the vibration modes. In the steady state (if exists) when energy supply is balanced by energy losses spectral energy transfer can be described by the Manley–Rowe power relations [107,108] (originally derived for classical macroscopic electrical oscillators [109]). The energy transfer may result in energy accumulation in some vibration mode(s) or excitation of solitons or soliton like waves. Therefore, energy supplied to such systems is not immediately thermalized, disturbs the thermodynamic equilibrium state and leads to energy condensation creating certain type of "order". This mechanism can exist in microscopic, mezoscopic, and macroscopic oscillators and vibration systems.

With no energy supply these systems sustain the thermodynamic equilibrium state and oscillations are random. If the energy supply is high the whole system may be far from thermodynamic equilibrium. Oscillations in excited mode(s) need not be random any longer. Nevertheless, energy transfer by a non-linear mechanism is random in nature. We will examine conditions for transition from randomness to coherence in non-linear systems with energy condensation. Before we start the analysis we have to mention that conditions for coherent oscillations are not included in the mechanism of spectral energy transfer and energy condensation. Energy condensation is only a necessary condition for creation of coherence.

First of all we will describe the mechanism of energy transfer and the concept of coherence. Energy transfer produced by non-linear mechanisms is random and represents a source of energy in the vibration modes accepting the energy. Energy is transferred between vibration modes in the mutually opposite directions and, therefore, certain feedback exists in the system. Energy transfer can result in the negative damping in the system (which is similar to the widely used concept of the negative resistance). Analyzing the effects of the negative damping we may find conditions for coherence in non-linear systems with energy condensation.

Creation of coherent states in vibration systems is a special problem of transition from disorder to order which is discussed and described e.g. in [36,37,41,42,132,133]. Coherence means a non-stochastic relation between the states (or processes) in arbitrary points x_1, t_1 and x_2, t_2 of the space-time domain. (A similar relation exists in the space-frequency domain too.) The correlation function and the autocorrelation function are used to express this relation, e.g. in [134,135]. In optics coherent and incoherent processes are defined as correlated and uncorrelated, respectively [134]. (Coherent and correlated processes are assumed to be equivalent.) This definition is connected with interference phenomena. It is coherence in the wide sense. We want to find conditions for periodic oscillations (e.g. for harmonic or quasi-harmonic oscillations). Therefore, we will use coherence in the narrow sense, i.e. for processes with periodic oscillations that have non-stochastic amplitudes and phases. Other correlated processes will be assumed to be incoherent.

We will analyze a system of interacting oscillators or a system of vibration modes which are equivalent to oscillators. Non-linearity may be an intrinsic property of the system or may be introduced by non-linear interactions with the surroundings. In polar systems these interactions can be mediated by the generated electric field (e.g in systems with polarization waves).

A harmonic oscillator with the noise may be described by a stochastic differential equation given e.g. in [136]. We will use the stochastic differential equation describing oscillation system with random excitation. This equation can be written in the form

$$\ddot{y}(t) + \beta\dot{y}(t) - \gamma\dot{y}(t) + \omega_0^2 y(t) = x(t) \tag{6.1}$$

where $x(t)$ is the input stationary stochastic process, $y(t)$ is the output process, ω_0 is the frequency of the system in question, β stands for the damping, and γ stands for the negative damping (representing the negative resistance effects). We note that ω_0, β, and γ are constants of the oscillation system. The term representing the negative resistance in Eq. (6.1) is connected with the output process (i.e. with $\dot{y}(t)$) and, therefore, depends on the positive feedback in the system. The systems have to be excited by energy supply. To solve the problem of coherence for a non-linear system with energy supply we have to find a solution of Eq. (6.1) with non-linear coefficients β and γ, i.e. with $\beta = \beta(y)$, $\gamma = \gamma(y)$. In this case the solution may be found in a special case or by numerical methods. We will assume that within certain range of excitation the positive damping β and the negative damping γ are constants. Therefore, the non-linear properties are represented by a constant value of the negative resistance. Confining our analysis to this simple case we are able to find a solution in the analytical form and to determine regions of incoherent and coherent behaviour. We have to emphasize here that solution of the non-linear stochastic equation might reveal new fundamental features (such as bifurcations or jumps between different branches of the solution).

6.2 Stochastic Differential Equation

Using the system function (the transfer function) in the form [136]

$$F(\mathrm{i}\omega) = \frac{1}{(\mathrm{i}\omega)^2 + (\beta - \gamma)\mathrm{i}\omega + \omega_0^2} \tag{6.2}$$

we can determine the output power spectrum $S_{yy}(\omega)$ for the input power spectrum $S_{xx}(\omega)$ from the relation

$$S_{yy}(\omega) = \frac{S_{xx}(\omega)}{|(\mathrm{i}\omega)^2 + (\beta - \gamma)\mathrm{i}\omega + \omega_0^2|^2} = \frac{S_{xx}(\omega)}{(\omega^2 - \omega_0^2)^2 + (\beta - \gamma)^2\omega^2} \,. \tag{6.3}$$

The input power spectrum is given by Planck's relation [137]. We will use it in the form

$$S_{xx}(\omega) = n\kappa \frac{\hbar\omega}{\exp\left(\dfrac{\hbar\omega}{kT}\right) - 1} \tag{6.4}$$

where n is determined by the number of the degrees of freedom and κ is a constant which may depend on the interaction of the system with its surroundings. In classical approximation when $\hbar\omega \ll kT$ we get from Eq. (6.4)

$$S_{xx}(\omega) = n\kappa kT = \alpha\kappa, \quad \text{where} \quad \alpha = nkT \,. \tag{6.5}$$

Using the relation for autocorrelation

$$R_{yy}(\tau) = \frac{1}{2\pi}\int_{-\infty}^{+\infty} S_{yy}(\omega)\exp(\mathrm{i}\omega\tau)\mathrm{d}\omega$$

(where τ is the time difference between two points) we get from Eq. (6.3) the following solutions for $\tau > 0$ (which is similar to the solution given in [136])

a) $(\beta - \gamma) > 0$

$$R_{yy}(\tau) = \frac{\alpha\kappa}{2(\beta - \gamma)\omega_0^2}\exp\left(-\frac{\beta - \gamma}{2}\tau\right)\left[\cos(\omega_1\tau) + \frac{\beta - \gamma}{2\omega_1}\sin(\omega_1\tau)\right]$$
$$\text{for } (\beta - \gamma)^2 < 4\omega_0^2 \tag{6.6}$$

$$R_{yy}(\tau) = \frac{\alpha\kappa}{2(\beta-\gamma)\omega_0^2} \exp\left(-\frac{\beta-\gamma}{2}\tau\right)\left[1+\frac{\beta-\gamma}{2}\tau\right]$$
$$\text{for } (\beta-\gamma)^2 = 4\omega_0^2 \qquad (6.7)$$

$$R_{yy}(\tau) = \frac{\alpha\kappa}{2(\beta-\gamma)\omega_0^2} \exp\left(-\frac{\beta-\gamma}{2}\tau\right)\left[\cosh(\omega_2\tau)+\frac{\beta-\gamma}{2\omega_2}\sinh(\omega_2\tau)\right]$$
$$\text{for } (\beta-\gamma)^2 > 4\omega_0^2 \qquad (6.8)$$

where $\omega_1 = \sqrt{\omega_0^2 - \left(\frac{\beta-\gamma}{2}\right)^2}$ and $\omega_2 = \sqrt{\left(\frac{\beta-\gamma}{2}\right)^2 - \omega_0^2}$.

For $(\beta-\gamma) \leq 0$ the autocorrelation of the output may be determined only in a certain time interval of finite length using analysis of transients. The method is described in [136]. We use the autocorrelation of the input stochastic process

$$\begin{aligned} R_{xx}(\tau) &= R_{xx}(t_1,t_2) = \alpha\kappa\delta(t_1-t_2) && \text{for } 0 \leq t_1,\ 0 \leq t_2, \text{ and } \tau = t_1 - t_2 \\ R_{xx}(\tau) &= R_{xx}(t_1,t_2) = 0 && \text{otherwise,} \end{aligned} \qquad (6.9)$$

the cross-correlation between the input and the output

$$R_{yx}(\tau) = R_{xy}(t_1,t_2) = \int_{-\infty}^{+\infty} R_{xx}(t_1-t',t_2)f(t')dt'$$

(where $f(t')$ is the impulse response) and the autocorrelation of the output

$$R_{yy}(\tau) = R_{yy}(t_1,t_2) = \int_{-\infty}^{+\infty} R_{yx}(t_1,t_2-t')f(t')dt'$$

where $0 \leq t_1$, $0 \leq t_2$, and $t_1 \geq t_2$. We assume that the impulse response is given by

$$f(t) = \frac{1}{\omega_0}\sin(\omega_0 t)U(t) \qquad \text{for } (\beta-\gamma) = 0$$
$$f(t) = \frac{1}{\omega_1}\exp\left(\frac{|\beta-\gamma|}{2}t\right)\sin(\omega_1 t)U(t) \qquad \text{for } (\beta-\gamma) < 0$$

where $U(t)$ is the unit step function ($U(t) = 1$ for $0 \leq t$, $U(t) = 0$ for $t < 0$).

The results are given for $\tau = t_1 - t_2 \geq 0$ in the form

b) $(\beta - \gamma) = 0$

$$R_{yy}(\tau) = R_{yy}(t_1, t_2)$$
$$= \frac{\alpha\kappa}{2\omega_0^2}\left\{\sin(\omega_0\tau)\frac{\sin^2(\omega_0 t_2)}{\omega_0} + \cos(\omega_0\tau)\left[t_2 - \frac{\cos(\omega_0 t_2)\sin(\omega_0 t_2)}{\omega_0}\right]\right\} \tag{6.10}$$

c) $(\beta - \gamma) < 0$

$$R_{yy}(\tau) = R_{yy}(t_1, t_2)$$
$$= \frac{\alpha\kappa}{\omega_1^2}\exp\left(\frac{|\beta-\gamma|}{2}\tau\right)\{\sin(\omega_1\tau)\rho_s(t_2) + \cos(\omega_1\tau)\rho_c(t_2)\} \tag{6.11}$$

where $|\beta - \gamma|$ is assumed to be small. The functions $\rho_s(t_2)$ and $\rho_c(t_2)$ in Eq. (6.11) are given by

$$\rho_s(t_2) = \frac{1}{4}\frac{\omega_1}{\omega_0^2}\left\{\exp(|\beta-\gamma|t_2)\left[\frac{|\beta-\gamma|}{2\omega_1}\sin(2\omega_1 t_2) - \cos 2\omega_1 t_2\right] + 1\right\}$$
$$\rho_c(t_2) = \frac{1}{2|\beta-\gamma|}\left[\exp(|\beta-\gamma|t_2) - 1\right]$$
$$-\frac{1}{8}\frac{|\beta-\gamma|}{\omega_0^2}\left\{\exp(|\beta-\gamma|t_2)\left[\frac{2\omega_1}{|\beta-\gamma|}\sin(2\omega_1 t_2) + \cos(2\omega_1 t_2)\right] - 1\right\} \tag{6.12}$$

The amplitudes of the functions $\rho_s(t_2)$ and $\rho_c(t_2)$ increase with t_2.

6.3 Coherence in Systems with Negative Resistance

Fig. 6.1a shows autocorrelation given by Eqs (6.6)–(6.8) normalized by $R_{yy}(0)$ of each curve. The amplitudes of the autocorrelation curves decrease with increasing time. The curve A given by Eq. (6.7) for $(\beta - \gamma)^2 = 4\omega_0^2$ denotes a boundary between incoherent states represented e.g. by the B curve and the time limited coherent states represented e.g. by the C and D curves (according to [134] both these cases should be denoted as the temporal coherent states). The curve A is the lower boundary of the time limited coherent states. The time limited coherence means that in any real system the oscillation state of the output process $y(t)$ can be maintained only over certain time interval which depends on the relaxation time $\Delta\tau$ of the exponentially decreasing amplitude of the autocorrelation. The smaller the value $\beta-\gamma$ the greater the relaxation time $\Delta\tau$.

For $(\beta-\gamma)=0$ a limit case of coherence occurs as follows from Eq. (6.10). The amplitude of $R_{yy}(\tau)$ does not depend on τ for a certain value of t_2 but increases linearly with increasing t_2 (on account of the term $t_2\cos(\omega_0\tau)$ on the right hand side of Eq. (6.10)). We refer to this state as the state of the "complete" coherence. We do not analyze here the resistance of the state to small fluctuations. In real physical systems this state is confined to a short time interval. When the amplitude of oscillations increases to a large value some non-linear mechanism interrupts the growth.

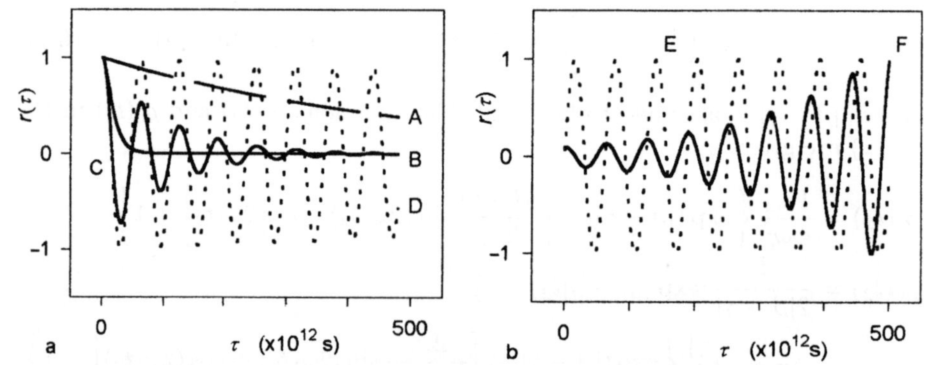

Fig. 6.1 The normalized autocorrelation $r_{yy}(\tau)$ versus the time difference τ. The parameter is the damping $(\beta-\gamma)$ of the vibration system.

a) Incoherent states (B) and time limited coherent states (C, D) – $(\beta-\gamma)>0$. The A curve is a boundary between incoherent and time limited coherent states.

b) "Complete" coherent state – $(\beta-\gamma)=0$ (E) and "super" coherent state – $(\beta-\gamma)<0$ (F).

Transition from the time limited coherence at very small positive values of $(\beta-\gamma)$ to the "complete" coherence ($(\beta-\gamma)=0$) is not described by the derived relations. (The time limited coherence is described by Eq. (6.6) and the "complete" coherence by Eq. (6.10).) We cannot assess properties of the system in this region with singularity. Nonetheless, in any case certain boundary $\epsilon\le(\beta-\gamma)$ (where $0<\epsilon$ is a small number) should exist up to which $R_{yy}(\tau)$ behaves "well". The state for $(\beta-\gamma)=\epsilon$ and $\exp\left(-\frac{1}{2}\epsilon\tau\right)\approx 1$ for a reasonably large τ forms – for practical purposes – the upper boundary of the time limited coherence in real systems.

Eq. (6.11) describes behaviour of the system for $(\beta-\gamma)<0$. In this case amplitude of autocorrelation exponentially increases with τ. This region may be denoted as a region of the "super" coherence. We note that this type of coherence can exist only within a limited time interval as the condition $\beta - \gamma < 0$ is satisfied in the time of a transient phenomenon.

Fig. 6.1b shows autocorrelation normalized by the value of $R_{yy}(\tau)$ at $\tau = 500\,\mathrm{ps}$ for the "complete" coherence ($(\beta - \gamma) = 0$) and for the "super" coherence ($(\beta - \gamma) < 0$) determined from Eq. (6.10)) and from Eq. (6.11), respectively. The normalized autocorrelation of the "complete" coherence is given by the E curve and that of the "super" coherence by the F curve.

The amplitudes of the autocorrelation of the time limited coherence given by Eqs (6.6)–(6.8) depend on the ratio $\kappa/[(\beta-\gamma)\omega_0^2]$ which is shown in Fig. 6.2 as a function of γ/β. The amplitude of the autocorrelation increases if $\beta - \gamma$ decreases. The quantity γ is determined by the energy condensation term in a non-linear system or in a system with non-linear interactions with the surroundings. Depending on energy supply to the system as well as on energy transfer between oscillators or between vibration modes the oscillation system may be shifted from the incoherent state to the time limited coherent state then further to the "complete" coherent state and to the "super" coherent state. We have to emphasize here that the time limited coherence may exist even with no negative damping if $4\omega_0^2 > \beta^2$ but the "complete" coherence and the "super" coherence depend on the energy supply to the system in question. Therefore, the "complete" coherence and the "super" coherence are connected with excitation above thermodynamic equilibrium of the system.

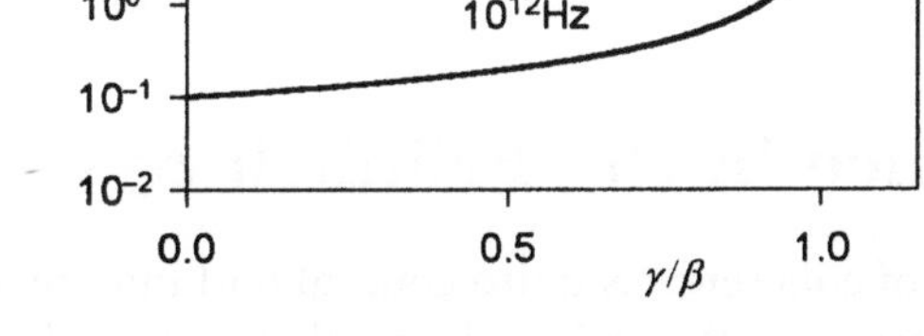

Fig. 6.2 Relative values of the amplitude factor $A = \kappa/[(\beta - \gamma)\omega_0^2]$ of the autocorrelation versus the ratio $\gamma/\beta < 1$. The parameter is the angular frequency ω_0 (10^{11} Hz and 10^{12} Hz).

Random excitation by itself without the negative damping cannot generate a "complete" coherent state. If the input power of the process $x(t)$ on the right hand side of Eq. (6.1) is increased only the amplitude α of the autocorrelation function given by the Eqs (6.6)–(6.8) is changed. The character of the autocorrelation function (and, therefore, behaviour of the system) remains unchanged and no transition from the incoherent state to the time limited coherent state and to the "complete" coherent state takes place.

The input process $x(t)$ in Eq. (6.1) may contain a deterministic signal with a spectrum given e.g. by the relation

$$S_{xx\mathrm{d}}(\omega) = A_\mathrm{d}\left[\delta(\omega - \omega_i) + \delta(\omega + \omega_i)\right] \tag{6.13}$$

where A_d is the amplitude. The autocorrelation function is given by

$$\begin{aligned} R_{yy\mathrm{d}}(\tau) &= \frac{1}{2\pi}\int_{-\infty}^{+\infty} A_\mathrm{d}\frac{\delta(\omega - \omega_i) + \delta(\omega + \omega_i)}{(\omega^2 - \omega_0^2)^2 + (\beta - \gamma)^2\omega^2}\exp(\mathrm{i}\omega\tau)\mathrm{d}\omega \\ &= \frac{1}{\pi}A_\mathrm{d}\frac{\cos(\omega_i\tau)}{(\omega_i^2 - \omega_0^2)^2 + (\beta - \gamma)^2\omega_i^2}\ . \end{aligned} \tag{6.14}$$

In this case the autocorrelation function always has besides a component $R_{yy}(\tau)$ an oscillating component $R_{yy\mathrm{d}}(\tau)$ given by Eq. (6.14). The oscillating component described by the autocorrelation $R_{yy\mathrm{d}}(\tau)$ has a constant amplitude and we assume that the system behaviour has certain similarity to a system which has zero damping, i.e. to a system where $(\beta - \gamma) = 0$.

Autocorrelation of a harmonic component with the frequency ω might be approximated by the relation

$$R(\tau) = \alpha_\mathrm{s}\exp(\mathrm{i}\omega\tau) \tag{6.15}$$

where α_s is the expectation value of the square of the amplitude of the harmonic component. Besides the harmonic component the signal may contain the noise with a similar expectation value α_n. The ratio of α_s to α_n may be used to assess the degree of coherence of the signal. The degree of coherence may be described by the autocorrelation coefficient too.

6.4 Coherence in the Fröhlich System

The derived theory of coherence is quite general and may be applied to a wide range of vibration systems. We will analyze coherence in the Fröhlich system.

In the Fröhlich system energy transfer between vibration modes is mediated by the heat bath. Therefore, the heat bath participates in the energy transfer. The system is linear and the non-linear properties are introduced through interaction with the heat bath. As a result of the spectral transformation energy is condensed in the lowest frequency mode and for sufficiently high energy supply the vibrations in this mode are assumed to be coherent e.g. in [7,8,75,79,81,138]. This mechanism can exist e.g. in systems with polarization waves where the generated "near zone" oscillating electric field mediates interactions with the surroundings and energy flow from the systems to the surroundings (to the heat bath) and back into the system. The non-linear term in the rate equation is given by the relation (5.10). This term is capable of creation of the negative damping γ due to energy transfer between oscillators or vibration modes. It depends on the excitation of vibration modes, i.e. on the occupation numbers. The $(\dot{n}_i)_\chi$ value may be positive, negative, or zero which means the negative damping, the positive damping, and no contribution to the damping, respectively. We have to emphasize that the term given by Eq. (5.10) represents a feedback in the system. Energy flows between the modes in both directions: from the i-th to the j-th mode and vice versa.

The linear term $(\dot{n}_i)_\Phi$ in the rate equation given by Eq. (5.5) represents the ordinary damping which is related to β in the stochastic differential equation. (Nevertheless, the $(\dot{n}_i)_\chi$ term contributes to β too.) $(\dot{n}_i)_\Phi$ in Eq. (5.5) depends on the occupation number of the vibration mode. This term increases with increasing n_i. The rate equation representing power balance in the i-th oscillator (or vibration mode) is given by Eq. (5.11) which can be written in the form

$$\dot{n}_i = s_i + (\dot{n}_i)_\Phi + (\dot{n}_i)_\chi \,.$$

In the transient time when $\dot{n}_i > s_i$ the total damping of the i-th vibration mode is negative, i.e. $(\beta - \gamma) < 0$. In the steady state $\dot{n}_i = 0$ and the Fröhlich system is stable as is proved by the Lyapunov method. The stability means that parameters of the system do not alter and regardless of the concrete dependence of β and γ on the occupation numbers these parameters remain constant. The energy supply to the system is balanced by the terms $(\dot{n}_i)_\Phi + (\dot{n}_i)_\chi$ which cause the positive damping, i.e. $0 < (\beta - \gamma)$. As $0 < (\beta - \gamma)$ then the Fröhlich system remains in the state of the limited coherence. Nevertheless, the difference $(\beta - \gamma)$ may be very small and coherence may be very high (as follows from the high amplitude factor A shown in Fig. 6.2) for a considerably long time.

7. Spectral Transformation of Energy

7.1 Manley–Rowe Power Relations

The Fröhlich rate equation describes the basic principle of energy condensation based on energy transfer between normal modes with different frequencies. This energy transfer depends on the non-linear interactions between the vibration system and the heat bath which represent two quanta processes. Energy can condense in the vibration states in proteins carrying the "static" electric dipoles or multipoles in the plasma membrane or inside a cell, e.g. in the cytoskeleton. The intensity of the electric field in the membrane is of the order of $10^5\,\mathrm{Vcm}^{-1}$. The charge distributions in proteins give rise to an electric field whose intensity inside the protein molecule may exceed the intensity of the electric field in the membrane. The intensity of the static electric field in protein molecules may be as high as $10^6\,\mathrm{Vcm}^{-1}$. (In a conductive liquid medium the mobile ions are attracted to the charged regions of proteins and screen the static electric field.) Some vibration systems in proteins are very likely in the non-linear region (the anharmonic vibrations). Therefore, energy transfer and energy condensation caused by an intrinsic non-linear mechanism may play an important role in proteins and polymer structures composed of proteins that have fundamental function in living cells.

When the energy supply to the system in question is high enough energy stored in the lowest frequency mode steeply increases. The lowest frequency mode is highly excited and its energy is much greater than the energy in thermodynamic equilibrium. A state far from thermodynamic equilibrium is achieved. Nevertheless, the potential energy profile need not correspond to harmonic vibrations. The energy condensed in the system may shift the whole vibration system to the non-linear region. Besides the original terms of the Fröhlich rate equation the terms representing the general multiple quanta processes may be important. The multiple quanta processes with participation of the heat bath as well as the processes with no participation of the heat bath may exist.

In Chapter 4 we derived the rate equations for the multiple quanta processes in non-linear systems. The rate equation was derived as a relation for expectation values using the Hamiltonian of the Wu–Austin type. In this section we will use the macroscopic approach to the problem. Macroscopic treatment of the non-linear spectral energy transformation is less complicated than the quantum mechanical one.

We will analyze the processes with no participation of the heat bath. We assume that the stationary conditions exist, i.e. that not only the energy of the whole vibration system is constant but also the energy of all normal modes with different frequencies. Energy transfer from a particular normal mode to all other normal modes and energy losses from it are balanced by the energy transferred to this mode from the other modes and by the energy supply. We will find a solution under the steady state conditions. The Fourier expansion technique will be used. We will analyze a simple model of a vibration system with two fundamental angular frequencies ω_1 and ω_2. If the vibrations are anharmonic the combination frequencies $m\omega_1 + n\omega_2$ (where m and n are the integers) may be generated. (In a real vibration system only certain combination frequencies are important which depends on the non-linear properties of the system.) We express the displacement q from the equilibrium position, the velocity v, and the force f of the vibration system in the form [107]

$$q = \sum_{m=-\infty}^{+\infty} \sum_{n=-\infty}^{+\infty} Q_{m,n} \exp\left[\mathrm{i}(m\omega_1 + n\omega_2)t\right] \tag{7.1}$$

$$v = \sum_{m=-\infty}^{+\infty} \sum_{n=-\infty}^{+\infty} V_{m,n} \exp\left[\mathrm{i}(m\omega_1 + n\omega_2)t\right] \tag{7.2}$$

$$f = \sum_{m=-\infty}^{+\infty} \sum_{n=-\infty}^{+\infty} F_{m,n} \exp\left[\mathrm{i}(m\omega_1 + n\omega_2)t\right] \tag{7.3}$$

where

$$Q_{m,n} = Q^*_{-m,-n}\,, \quad V_{m,n} = V^*_{-m,-n}\,, \quad \text{and} \quad F_{m,n} = F^*_{-m,-n} \tag{7.4}$$

are the complex amplitudes (the coefficients of the Fourier expansion) of the displacement, of the velocity, and of the force, respectively. We have to emphasize that Eqs (7.1)–(7.3) describe the steady state of a vibration system and that the energy in the vibration modes does not depend on time.

The $F_{m,n}$ amplitudes are given by

$$F_{m,n} = \frac{1}{(2\pi)^2} \int_0^{2\pi} \mathrm{d}\eta \int_0^{2\pi} f \exp\left[-\mathrm{i}\left(m\xi + n\eta\right)\right] \mathrm{d}\xi \tag{7.5}$$

where $\xi = \omega_1 t$, $\eta = \omega_2 t$. Let us multiply both sides of Eq. (7.5) by $\mathrm{i}mQ^*_{m,n}$ and sum from $-\infty$ to $+\infty$ over m and n. We change the order of integration and summation on the right hand side of Eq. (7.5) and get

$$\sum_{m=-\infty}^{+\infty} \sum_{n=-\infty}^{+\infty} imF_{m,n}Q^*_{m,n}$$
$$= \frac{1}{(2\pi)^2} \int_0^{2\pi} d\eta \int_0^{2\pi} f \sum_{m=-\infty}^{+\infty} \sum_{n=-\infty}^{+\infty} imQ^*_{m,n} \exp\left[-i(m\xi + n\eta)\right] d\xi \, . \tag{7.6}$$

After substitution of certain expressions derived from Eqs (7.1) and (7.2) and some arrangements Eq. (7.6) will yield the final result. We will describe the substitutions and arrangements. Differentiating Eq. (7.1) with respect to $\xi = \omega_1 t$ and using Eq. (7.4) we get

$$\frac{\partial q}{\partial \xi} = - \sum_{m=-\infty}^{+\infty} \sum_{n=-\infty}^{+\infty} imQ^*_{m,n} \exp\left[-i(m\xi + n\eta)\right] \tag{7.7}$$

and from Eqs (7.1) and (7.2) we may derive

$$v = \frac{dq}{dt} = - \sum_{m=-\infty}^{+\infty} \sum_{n=-\infty}^{+\infty} i\left(m\omega_1 + n\omega_2\right) Q^*_{m,n} \exp\left[-i(m\omega_1 + n\omega_2)t\right]$$
$$= \sum_{m=-\infty}^{+\infty} \sum_{n=-\infty}^{+\infty} V^*_{m,n} \exp\left[-i(m\omega_1 + n\omega_2)t\right] \tag{7.8}$$

and further

$$Q^*_{m,n} = i \frac{V^*_{m,n}}{m\omega_1 + n\omega_2} \, . \tag{7.9}$$

Substituting $\partial q/\partial \xi$ from Eq. (7.7) for the double summation term inside the integral on the right hand side of Eq. (7.6) and Eq. (7.9) for $Q^*_{m,n}$ on the left hand side of the same equation one gets

$$\sum_{m=-\infty}^{+\infty} \sum_{n=-\infty}^{+\infty} m \frac{F_{m,n}V^*_{m,n}}{m\omega_1 + n\omega_2} = \frac{1}{(2\pi)^2} \int_0^{2\pi} d\eta \int_0^{2\pi} f \frac{\partial q}{\partial \xi} d\xi \, . \tag{7.10}$$

The left hand side of Eq. (7.10) may be arranged using the relation

$$P_{m,n} = \frac{1}{T_{m,n}} \int_0^{T_{m,n}} P_{m,n}\left(t\right) dt = F^*_{m,n}V_{m,n} + F_{m,n}V^*_{m,n} \tag{7.11}$$

for the time average $P_{m,n}$ of the power of the harmonic component with frequency $m\omega_1 + n\omega_2$. $T_{m,n} = 2\pi/(m\omega_1 + n\omega_2)$ in Eq. (7.11) is the period of oscillations. The arrangement of the relation on the left hand side of Eq. (7.10) is performed in the following way

$$\sum_{m=-\infty}^{+\infty}\sum_{n=-\infty}^{+\infty} m\frac{F_{m,n}V^*_{m,n}}{m\omega_1+n\omega_2} = \sum_{n=-\infty}^{+\infty}\left(\sum_{m=-\infty}^{0} m\frac{F_{m,n}V^*_{m,n}}{m\omega_1+n\omega_2} + \sum_{m=0}^{+\infty} m\frac{F_{m,n}V^*_{m,n}}{m\omega_1+n\omega_2}\right) . \tag{7.12}$$

Using Eq. (7.4) the first term on the right hand side of Eq. (7.12) gives

$$\sum_{n=-\infty}^{+\infty}\sum_{m=-\infty}^{0} m\frac{F_{m,n}V^*_{m,n}}{m\omega_1+n\omega_2} = \sum_{m=0}^{+\infty}\sum_{n=+\infty}^{-\infty} -m\frac{F_{-m,-n}V^*_{-m,-n}}{-m\omega_1-n\omega_2} = \sum_{m=0}^{+\infty}\sum_{n=-\infty}^{+\infty} m\frac{F^*_{m,n}V_{m,n}}{m\omega_1+n\omega_2} \tag{7.13}$$

and Eq. (7.12) may be written in the form

$$\sum_{m=-\infty}^{+\infty}\sum_{n=-\infty}^{+\infty} m\frac{F_{m,n}V^*_{m,n}}{m\omega_1+n\omega_2} = \sum_{m=0}^{+\infty}\sum_{n=-\infty}^{+\infty} m\frac{F^*_{m,n}V_{m,n}+F_{m,n}V^*_{m,n}}{m\omega_1+n\omega_2} = \sum_{m=0}^{+\infty}\sum_{n=-\infty}^{+\infty} m\frac{P_{m,n}}{m\omega_1+n\omega_2} \tag{7.14}$$

where the average power $P_{m,n}$ at the frequency $m\omega_1 + n\omega_2$ is given by Eq. (7.11). Eq. (7.10) has the final form

$$\sum_{m=0}^{+\infty}\sum_{n=-\infty}^{+\infty} m\frac{P_{m,n}}{\omega_1+n\omega_2} = \frac{1}{(2\pi)^2}\int_0^{2\pi} \mathrm{d}\eta \int_0^{2\pi} f\frac{\partial q}{\partial \xi}\mathrm{d}\xi . \tag{7.15}$$

A similar analysis may be performed in which the roles of ξ and η are interchanged. The result is

$$\sum_{n=0}^{+\infty}\sum_{m=-\infty}^{+\infty} n\frac{P_{m,n}}{m\omega_1+n\omega_2} = \frac{1}{(2\pi)^2}\int_0^{2\pi} \mathrm{d}\xi \int_0^{2\pi} f\frac{\partial q}{\partial \eta}\mathrm{d}\eta . \tag{7.16}$$

The right hand sides of Eqs (7.15) and (7.16) need closer inspection. If $\partial q/\partial\xi$ in Eq. (7.15) is antisymmetric with respect to the maximum value of $q = q_{\max}$ (or to the minimum value $q = -q_{\max}$) and f is symmetric with respect to this point the right hand side of Eq. (7.15) is equal to zero. This condition is satisfied e.g. for an oscillator with no losses of energy if both directions of motion of the oscillating mass are equivalent. The same holds for Eq. (7.16) where $\partial q/\partial\eta$ is instead of $\partial q/\partial\xi$. Figs 7.1a and 7.1b show the absolute value of the force acting in the oscillator as a function of the displacement and the velocity versus time, respectively. The point $q = q_{\max}$ ($q = -q_{\max}$) corresponds to the time $t = T/4$ ($t = 3T/4$) in Fig. 7.1b. The force and the velocity satisfy the condition of symmetry and antisymmetry, respectively. The velocity $v = \mathrm{d}q/\mathrm{d}t$ differs from $\partial q/\partial\xi$ only by the amplitudes of the harmonic terms (as follows from Eq. (7.1). The antisymmetry with respect to the point $q_{\max}$ (or $-q_{\max}$) remains unchanged.

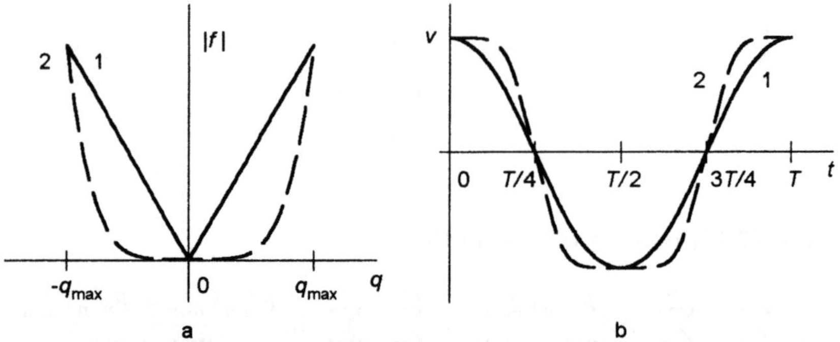

Fig. 7.1 Absolute value of the force $|f|$ as a function of the displacement q from the equilibrium position $q = 0$ (a) and the velocity v as a function of the time t (b) for a linear and a non-linear oscillator (denoted by 1 and 2, respectively).

Eqs (7.15) and (7.16) become

$$\sum_{m=0}^{+\infty}\sum_{n=-\infty}^{+\infty} m\frac{P_{m,n}}{m\omega_1 + n\omega_2} = 0 \tag{7.17}$$

$$\sum_{n=0}^{+\infty}\sum_{m=-\infty}^{+\infty} n\frac{P_{m,n}}{m\omega_1 + n\omega_2} = 0\,. \tag{7.18}$$

Eqs (7.17) and (7.18) are the well known Manley–Rowe relations. The Manley–Rowe relations were derived for classical non-linear electric oscillators in [109] and for mechanical oscillators and quantum mechanical oscillators in [107,108].

In Eqs (7.17) and (7.18) the power $P_{m,n}$ is related to the angular frequency given in the form $m\omega_1 + n\omega_2$. This frequency is a combination of two fundamental frequencies ω_1 and ω_2. The combination frequencies $m\omega_1 + n\omega_2$ are created on account of the non-linear properties of the oscillator. Generalization of Eqs (7.17) and (7.18) for more fundamental frequencies may be easily done.

Eqs (7.17) and (7.18) determine the energy transfer between the modes with convenient frequency relation by a non-linear mechanism. The direction of energy transfer may be determined only relatively in special cases (in systems with two or three Fourier components), i.e. the power flow in one Fourier component has opposite direction that the flow in the other components. The power relations determined as the expectation values from the Hamiltonian (in Chapter 4) enable us to determine the direction of energy transfer in each mode. The direction of energy transfer depends on the occupation numbers of particular modes.

Parametric principle of energy transfer was widely used in engineering and theoretical analysis may be found e.g. in [139–144].

7.2 Energy Transfer Between Modes

We accept a simple model of three normal modes oscillating at the frequencies $\omega_{m,n} = m\omega_1 + n\omega_2$ for

$$\begin{aligned} m &= 1\,,\ n = 0\,, &\quad \text{i.e.} \quad \omega_{1,0} &= \omega_1 \\ m &= 0\,,\ n = 1\,, &\quad \text{i.e.} \quad \omega_{0,1} &= \omega_2 \\ m &= 1\,,\ n = 1\,, &\quad \text{i.e.} \quad \omega_{1,1} &= \omega_1 + \omega_2\,. \end{aligned}$$

The values of power corresponding to the particular modes with the frequencies $\omega_{1,0}$, $\omega_{0,1}$, and $\omega_{1,1}$ are $P_{1,0}$, $P_{0,1}$, and $P_{1,1}$, respectively. From Eqs (7.17) and (7.18) one obtains

$$\frac{P_{1,0}}{\omega_1} + \frac{P_{1,1}}{\omega_1 + \omega_2} = 0\,, \qquad \frac{P_{0,1}}{\omega_2} + \frac{P_{1,1}}{\omega_1 + \omega_2} = 0\,. \tag{7.19}$$

From Eqs (7.19) we get

$$\frac{P_{1,0}}{\omega_1} - \frac{P_{0,1}}{\omega_2} = 0\,. \tag{7.20}$$

If both the values $P_{1,0}$, $P_{0,1}$ are positive the power $P_{1,1}$ is negative and vice versa as follows from Eq. (7.20). It means that energy simultaneously flows from the ω_1 mode and from the ω_2 mode to the $(\omega_1 + \omega_2)$ mode (or vice versa). The direction of energy flow depends on the energy condensed in particular modes. In thermodynamic equilibrium the resulting energy flow between the modes vanishes. A similar stationary state might exist even far from thermodynamic equilibrium. Eqs (7.17) and (7.18) do not contain information about the stationary state and the direction of the flow of energy between the vibration modes may be assessed only relatively.

The direction of flow may be obtained from Eqs (4.56) and (4.57). The flow of energy depends on the quantity B. In this case B is proportional to

$$B \simeq (n_{1,1}+1)n_{1,0}n_{0,1} - n_{1,1}(n_{1,0}+1)(n_{0,1}+1) \tag{7.21}$$

where $n_{1,0}$, $n_{0,1}$, and $n_{1,1}$ are the occupation numbers of the modes with the frequencies $\omega_{1,0}$, $\omega_{0,1}$, and $\omega_{1,1}$, respectively. The right hand side of the relation (7.21) equals zero in thermodynamic equilibrium. The values $P_{1,0}$, $P_{0,1}$, and $P_{1,1}$ are proportional to B given by Eq. (4.57), i.e.

$$P_{1,0} \simeq -\omega_{1,0}B\,, \qquad P_{0,1} \simeq -\omega_{0,1}B\,, \qquad P_{1,1} \simeq +\omega_{1,1}B\,. \tag{7.22}$$

If the mode with the frequency $\omega_{0,1}$ is highly excited, i.e. $n_{0,1} \gg n_{1,0}$, $n_{0,1} \gg n_{1,1}$, and $n_{1,0} > n_{1,1}$ (we may derive the exact condition $n_{0,1} > (n_{1,0}++1)n_{1,1}/(n_{1,0}-n_{1,1})$ from Eq. (7.21) for $B > 0$) energy is transferred upwards from the modes $\omega_{1,0}$ and $\omega_{0,1}$ to the mode $\omega_{1,1}$. From Eq. (7.19) we determine the ratio of the output power $P_{1,1}$ to the input power $P_{1,0}$ in the form

$$\frac{-P_{1,1}}{P_{1,0}} = \frac{\omega_{1,1}}{\omega_1} = \frac{\omega_1+\omega_2}{\omega_1}\,. \tag{7.23}$$

The system behaves as an amplifier. The output power $P_{1,1}$ corresponds to the amplified input power $P_{1,0}$. The power gain given by Eq. (7.23) is proportional to the ratio of the frequencies of the modes ($\omega_1+\omega_2$ divided by ω_1). We have to mention one special property of this type of amplifier. The output is not coupled to the input. The input power is independent of the output power. There is no feedback in the system. The system is not liable to oscillate.

If the energy condensed in the mode with the frequency $\omega_{1,1} = \omega_1 + \omega_2$ is high, i.e. $n_{1,1} \gg n_{1,0}$ and $n_{1,1} \gg n_{0,1}$ (we may derive the exact condition $n_{1,1} > n_{1,0}n_{0,1}/(n_{1,0}+n_{0,1}+1)$ from Eq. (7.21) for $B < 0$) energy is transferred from the mode with the frequency $\omega_{1,1}$ to the low frequency modes with the frequencies $\omega_{1,0}$ and $\omega_{0,1}$ as follows from Eq. (7.22). We assume that $\omega_{1,0} < \omega_{0,1} < \omega_{1,1}$. The system amplifies the input signal with the frequency $\omega_{1,0}$. From Eq. (7.20) we get the power gain by the ratio

$$\frac{P_{0,1}}{P_{1,0}} = \frac{\omega_{0,1}}{\omega_{1,0}} = \frac{\omega_2}{\omega_1}\,. \tag{7.24}$$

In this case energy is transferred from the excited mode with the frequency $\omega_{1,1}$ to the modes with the frequencies $\omega_{0,1}$ and $\omega_{1,0}$. The energy transferred to the mode with the frequency $\omega_{1,0}$ depends on the output energy in the mode with the frequency $\omega_{0,1}$. The transferred energy is amplified. The feedback in the system can lead to oscillations.

The lowest frequency mode of the Fröhlich system may be highly excited as a consequence of energy condensation. An external signal may cause vibrations in the Fröhlich system with the frequency smaller than the frequency of the excited mode. Due to amplification the amplitude of the original small vibrations caused by the external signal may considerably increase and even get over the threshold of the self-sustained oscillations.

7.3 Transfer of Energy in Fröhlich Systems

A non-linear medium subject to oscillations does not sustain the original spectrum; energy distribution in the spectral lines is altered and new spectral components are excited. In non-linear systems where a steady state can be established the changes of the spectrum may be limited to the transient period. (We remark that this state can exist only in certain systems.) Regardless of the non-linear mechanism spectral energy transfer is a fundamental property of the non-linear systems. The Fröhlich system behaves like a special type of the non-linear system. The Fröhlich system is linear by itself and the non-linear behaviour is introduced by the non-linear interaction with the heat bath. We will examine the spectral transformation in the Fröhlich system. Interaction between neighbouring spectral lines in various non-linear systems may be much stronger than interaction between lines that are more distant. This quasi-resonant energy transfer will be introduced by the assumption that energy is transferred only between the closest neighbours. Further we will assume that the system is excited at the time $t = 0$ and for $t > 0$ no energy supply exists. The Fröhlich rate equation may be written in the form

$$\begin{aligned}\dot{n}_i = &-\Phi_i\left[n_i \exp\left(\frac{h\nu}{kT}\right) - (n_i+1)\right]\\ &-\chi_{i,i-1}\left[n_i(n_{i-1}+1)\exp\left(\frac{h\nu_i}{kT}\right) - (n_i+1)n_{i-1}\exp\left(\frac{h\nu_{i-1}}{kT}\right)\right]\\ &-\chi_{i,i+1}\left[n_i(n_{i+1}+1)\exp\left(\frac{h\nu_i}{kT}\right) - (n_i+1)n_{i+1}\exp\left(\frac{h\nu_{i+1}}{kT}\right)\right]\,.\end{aligned} \tag{7.25}$$

We adopt a model of 30 normal modes in the frequency range from 10^{11} Hz to 10^{13} Hz; the modes are equidistant on the logarithmic scale. The transition probabilities $\Phi_i = 10^{-3}\,\mathrm{s}^{-1}$, $\chi_{i,i-1} = \chi_{i,i+1} = 10^{-4}\,\mathrm{s}^{-1}$, and the coefficient of

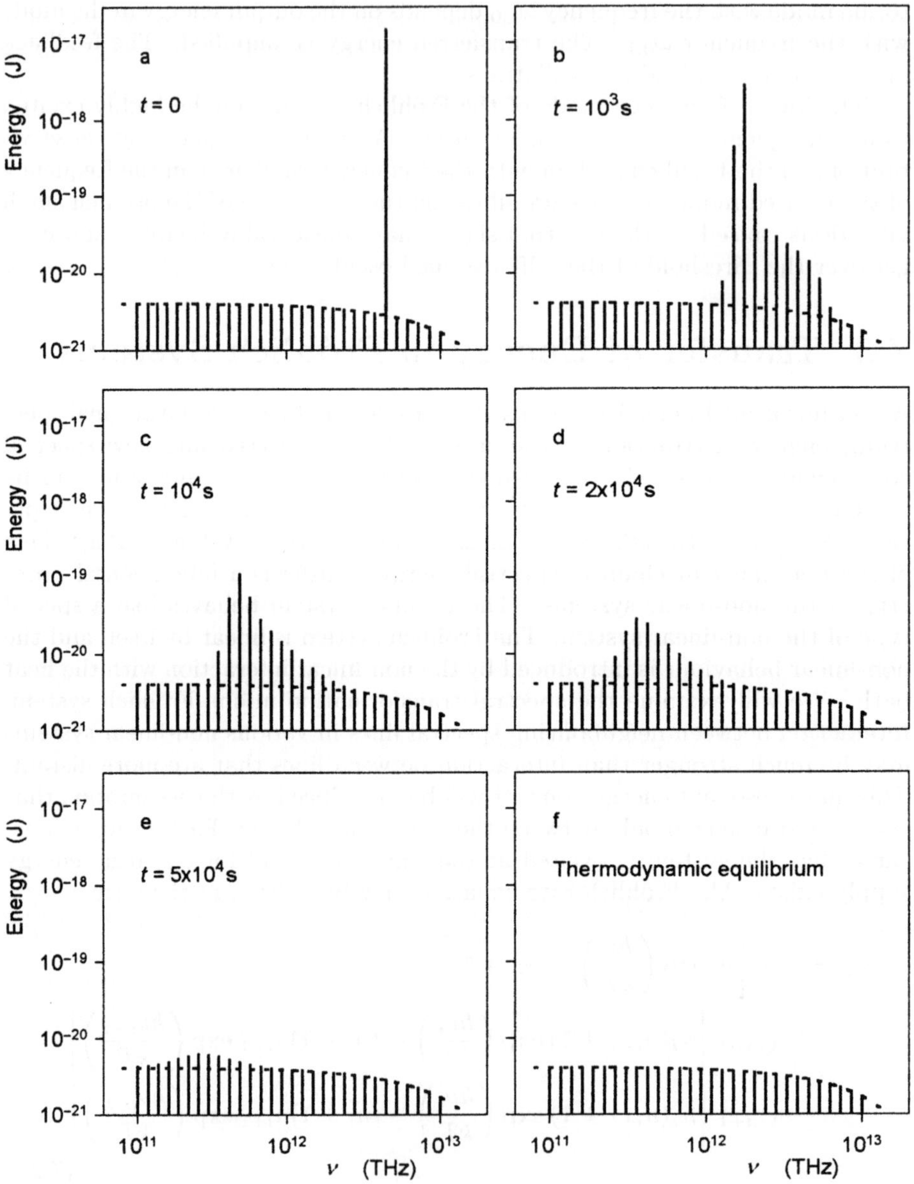

Fig. 7.2 Spectral energy transformation in a Fröhlich system. The evolution of the energy of the vibration modes versus frequency is shown for several discrete times. The system at the initial time $t = 0$ (a). The system at the time $t = 10^3$ s (b), $t = 10^4$ s (c), $t = 2 \times 10^4$ s (d), $t = 5 \times 10^4$ s (e), and in thermodynamic equilibrium (f). The dotted lines indicate the values of energy in thermodynamic equilibrium.

the energy supply $S_i = 0$ for all i. At the initial time the 25th mode with the frequency $\nu_{25} = 4.52\,\mathrm{THz}$ is highly excited ($n_{25} = 5000$). The other modes are in thermodynamic equilibrium (Fig. 7.2a). The time evolution of the system is shown in Fig. 7.2b–f. Energy is transferred predominantly to the low frequency modes. The energy transfer has a form of the wave in the spectral region travelling downwards along the frequency scale. The front edge of the wave is much steeper than the trailing edge. The spectral travelling wave is gradually attenuated; its energy decreases. Fig. 7.2f shows the state in thermodynamic equilibrium.

The model shown in Fig. 7.2 proves that the spectral transformation in the Fröhlich system favours the direction towards the low frequency modes. If the spectral energy transfer is not limited to the closest neighbours energy channelling downwards is faster and includes simultaneously all the low frequency modes in question.

7.4 Amplification of Electric Signals in Macromolecules

Macromolecules and molecular structures in the biological systems interact with an electric field. The oscillating electric field induces vibrations of atoms. We assume a one dimensional chain (similar to the chain shown in Fig. 3.1) with non-linear interactions between neigbouring atoms. Vibrations of an atom in the chain with the non-linear interactions between atoms may be described by the relation [145]

$$m\ddot{q}_n + \beta(2q_n - q_{n-1} - q_{n+1}) + \gamma[2q_n^2 + q_{n-1}^2 + q_{n+1}^2 - 2q_n(q_{n-1} + q_{n+1})] = 0 \tag{7.26}$$

where m is the mass of the atoms in the chain, q_n (q_{n-1}, q_{n+1}) is the displacement of the n-th (($n-1$)-th, ($n+1$)-th) atom from the equilibrium position ($\ddot{q}_n$ is the second derivative of q_n with respect to time), and β (γ) is the coefficient of linear (non-linear) interaction between the neighbouring atoms, respectively. The absolute value of the coefficient γ is supposed to be considerably smaller than the value of the coefficient β.

The macromolecules receive an external electromagnetic signal. Under convenient conditions the reception capabilities of macromolecules and structures may be connected with signal processing, i.e. the signal received is amplified by an internal mechanism. We present a simple theory of the amplification mechanism.

We assume that the frequencies and the absolute values of the wave vectors of vibrations in the chain satisfy the relations

$$\omega_1 + \omega_2 = \omega_3, \quad k_1 + k_2 = k_3 \tag{7.27}$$

where k_1, k_2, k_3 are the absolute values of the wave vectors. The steady state solution of Eq. (7.26) is assumed in the form

$$q_n = \sum_{j=1}^{3} \{A_j \exp[\mathrm{i}(k_j an - \omega_j t)] + A_{-j} \exp[-\mathrm{i}(k_j an - \omega_j t)]\} \tag{7.28}$$

where a is the distance between the atoms in the chain. We replace q_n in Eq. (7.26) by the relation (7.28). Similar substitutions are made for q_{n-1} and q_{n+1} in Eq. (7.26). The terms of each harmonic component equal zero independently of the other components. After elementary arrangements we use Eqs (7.27) and get for oscillations at the frequency ω_3 the relation [145]

$$\begin{aligned} &A_3 \left\{-\omega^2 m + \beta\,[2 - \exp(-\mathrm{i}k_3 a) - \exp(\mathrm{i}k_3 a)]\right\} + \\ &\quad +2\gamma A_1 A_2 \{2 + \exp[-\mathrm{i}(k_1 + k_2)a] + \exp[\mathrm{i}(k_1 + k_2)a] - \exp(-\mathrm{i}k_1 a) \\ &\quad - \exp(\mathrm{i}k_1 a) - \exp(-\mathrm{i}k_2 a) - \exp(\mathrm{i}k_2 a)\} = 0 \end{aligned} \tag{7.29}$$

which yields

$$A_3 = 4\gamma A_1 A_2 \frac{1 + \cos(k_3 a) - [\cos(k_1 a) + \cos(k_2 a)]}{m\omega_1^2 - 2\beta\,[1 - \cos(k_3 a)]} . \tag{7.30}$$

Similar relations hold for A_1 and A_2

$$A_1 = 4\gamma A_3 A_{-2} \frac{1 + \cos(k_1 a) - [\cos(k_2 a) + \cos(k_3 a)]}{m\omega_1^2 - 2\beta\,[1 - \cos(k_1 a)]} \tag{7.31}$$

$$A_2 = 4\gamma A_3 A_{-1} \frac{1 + \cos(k_2 a) - [\cos(k_1 a) + \cos(k_3 a)]}{m\omega_2^2 - 2\beta\,[1 - \cos(k_2 a)]} . \tag{7.32}$$

Eqs (7.30)–(7.32) reveal how the vibrations with the freqencies ω_1, ω_2, and ω_3 are mutually bound. For example, the amplitude A_2 in Eq. (7.32) is proportional to the amplitude A_{-1}. If the amplitude A_3 is large (which may be satisfied for a Fröhlich excited system) the signal at the frequency ω_1 is amplified and converted to the signal at the frequency ω_2. Energy is supplied from the mode ω_3. The system can also amplify external signals when the amplitude A_1 (or A_2) is large. This class of amplifiers is called the travelling wave amplifiers.

In Section 7.2 we discussed oscillators which have the same frequency relation (i.e. $\omega_3 = \omega_1 + \omega_2$). We found out that the input energy at the frequency ω_1 is amplified (with the power gain equal to ω_2/ω_1 given by Eq.

(7.24)) and the output is released at the frequency ω_2 and vice versa (energy at the frequency ω_2 is transferred downwards to the input). The system can be unstable. If the excitation of the mode ω_3 is sufficiently high the system can start to oscillate.

This mechanism of parametric amplification can explain the effects of weak external electromagnetic fields on biological systems. The macromolecules and the polymer structures containing chains of atoms can amplify external signals. The power of the output may be even several orders greater than the power of the input signal. The resulting biological effects may depend on the energy of the output signal. If the energy is greater than certain "critical" level then the corresponding biological effect is triggered. Amplification depends on the "pumping" signal supplying energy to the system which acts as an "amplifier". Therefore, amplification is conditioned by the excitation of the system. The energy supply may vary with the life cycle and with the physiological conditions. Therefore, we may expect that biological effects of the external electromagnetic fields depend on the excitation, i.e. on the internal state of the biological systems.

8. Oscillating Electric Field Generated by Living Cells

8.1 Source of the Oscillating Electric Field

We may assume that the coherent vibration states are excited in protein molecules in cellular membranes. The lipid bilayer of the membrane can oscillate too. Nevertheless, this is not the only possibility. Proteins inside a cell may be excited as well. In particular the cytoskeleton can play an important role in this respect. The microtubules are a fundamental part of the cytoskeleton and its main organizer. Vibrations in the microtubules can be excited by energy supply from GTP. Beside the microtubules the actin filaments have large supply of energy (from ATP).

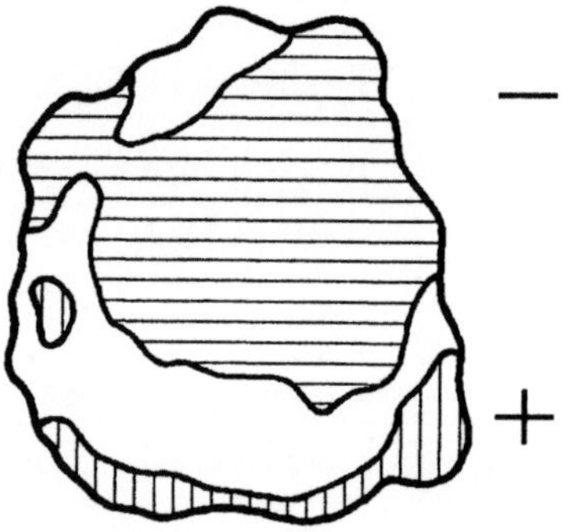

Fig. 8.1 Electrostatic potential distribution in a globular protein (oxidized flavodoxin). Only regions below about −13 V and above about −7 V (deno ted by − and by + signs, respectively) are shown. Computer simulation. After [119].

Electric potential and charge distribution in proteins were analyzed by Wada et al. in [116–119]. The static charge distribution in proteins has mainly the dipole, the quadrupole, and the octupole character. Fig. 8.1 shows a molecule of the globular protein – oxidized flavodoxin [119]. There are regions with the negative potential and regions with the positive potential. The protein charge distribution has a dipole structure. The giant breathing and twisting vibrations in biological molecules may have the dipole as well as the multipole character. The polar vibrations can have non-linear interactions with the heat bath.

The intensity of the electric field inside a protein molecule may be of about 10^5–10^6 V/cm. On account of the high values of the electric field in the protein molecules some of the vibrations are very likely anharmonic. Energy transfer between the normal modes with no participation of the heat bath may occur.

Excited proteins or structures of proteins represent oscillation centres. Some protein molecules can move in the membrane in the lateral direction without changing their orientations with respect to the membrane bilayer (i.e. without a flip-flop). Distribution of the protein oscillation centres in a membrane may depend on the electrostatic interaction as well as on the electromagnetic interaction. (We have to point out that the oscillation system in a cell may include the cytoskeleton; this may have a profound effect on space distribution of the oscillation centres.) The electrostatic forces might be especially effective among identical protein molecules and among molecules with similar charge distribution in a medium with low conductivity. The intracellular liquid medium as well as the intercellular medium are conductive. Ions and charged particles in the ambient medium are attracted to the charged regions of proteins. The protein charged regions are screened. Due to screening by mobile ions and charged particles the electrostatic interaction is a short range interaction (Appendix 13.13). The mobility and concentration of ions and charged particles in the cytosol and in the intercellular liquid medium determine the highest frequency of oscillation that can be efectively sreened. If the frequency of oscillations is high enough the ions remain at rest and have

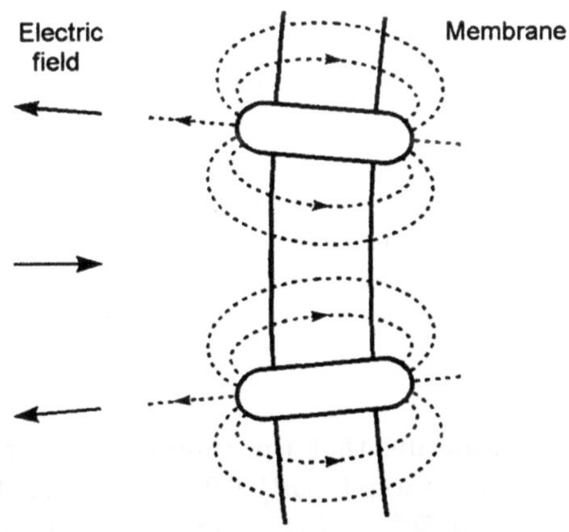

Fig. 8.2 A schematic picture of a part of the cellular membrane with the oscillation centres and with the line of force of the generated electric field. The dipole oscillators are assumed. In real systems the oscillation centres may be connected with the cytoskeleton. The arrows indicate the direction of the intensity of the electric field.

no effect on the generated electric field. The interaction forces mediated by the oscillating electric field may act upon the protein molecules, which form the oscillation centres. The resulting distribution of the oscillation centres has to satisfy a condition given by the principle of the minimum of the free energy of the whole system. The energy of oscillation centres together with the energy of the generated electromagnetic field have to be minimal. In the case of spherical cellular membrane embedded in a homogenous medium whose properties do not depend on space coordinates the distribution of the identical coherent oscillation centres should be uniform.

A schematic model of a small part of the cellular membrane with the oscillation centres and with the line of force of the intensity of the generated electric field is shown in Fig. 8.2. The interacting oscillation centres may create a unified vibration system of the cell. The vibrations in the cell have to be understood as a whole.

8.2 Continuous Dipole Layer

We will examine the electromagnetic field of a spherical model cell with oscillating homogenous dipole layer on its surface, i.e. of a macroscopic model with continuous dipole layer. The generated electromagnetic field may be evaluated from the solution of the vector wave equation [146]

$$\nabla^2 \boldsymbol{C} + k^2 \boldsymbol{C} = 0 \tag{8.1}$$

where $k^2 = \mu\varepsilon\omega^2 + \mathrm{i}\mu\sigma\omega$ is the propagation constant (μ is the permeability, ε is the permittivity, and σ is the conductivity) and $\boldsymbol{C}$ denotes any vector of the electromagnetic field. The field vectors of the intensity of the electric field $\boldsymbol{E}$, the magnetic filed $\boldsymbol{H}$, the electric displacement $\boldsymbol{D}$, the magnetic induction $\boldsymbol{B}$, the vector potential $\boldsymbol{A}$, and the Hertzian vectors $\boldsymbol{\Pi}$, $\boldsymbol{\Pi}^*$ satisfy one and the same equation. We will find a solution of Eq. (8.1) in a closed domain outside the spherical dipole layer. The medium in the domain is assumed to be homogeneous, isotropic, and with no sources. Solution of the vector wave equation in spherical coordinates may be derived from the characteristic functions of the corresponding scalar wave equation

$$\nabla^2 \Psi + k^2 \Psi = 0\,. \tag{8.2}$$

Assuming that the time dependent part of the scalar function Ψ may be written as $\exp(-\mathrm{i}\omega t)$ ($\Psi = \exp(-\mathrm{i}\omega t) f(R, \Theta, \Phi)$) and that the scalar wave equation $\nabla^2 f + k_2 f = 0$ is separable, i.e.

$$f = f_1(R) f_2(\Theta) f_3(\Phi)$$

the elementary solution (the characteristic solution) in spherical coordinates R, Θ, Φ my be given in the form [146]

$$f_{m,n\{\substack{\mathrm{e}\\ \mathrm{o}}\}} = z_n(kR) P_n^m(\cos\Theta) \left\{ \begin{array}{c} \cos(m\Phi) \\ \sin(m\Phi) \end{array} \right\} \tag{8.3}$$

where

$$z_n(kR) = \sqrt{\frac{\pi}{2kR}} Z_{n+\frac{1}{2}}(kR) \tag{8.4}$$

$$P_n^m(\cos\Theta) = (1-\cos^2\Theta)^{m/2}\frac{\mathrm{d}^m P_n(\cos\Theta)}{\mathrm{d}(\cos\Theta)^m} \tag{8.5}$$

are the spherical Bessel functions and the associated Legendre polynomials, respectively [146]. $Z_{n+\frac{1}{2}}(kR)$ in Eq. (8.4) stands for circular cylinder functions which may be the Bessel functions of the first kind, of the second kind, or of the third kind (the Hankel functions). The first three members of the associated Legendre polynomials family are given by the relations

$$P_0(\cos\Theta) = 1\,, \quad P_1(\cos\Theta) = \cos\Theta\,, \quad P_1^1(\cos\Theta) = \sin\Theta\,. \tag{8.6}$$

The general solution of Eq. (8.2) is given by a sum of the characteristic solutions given by Eq. (8.3) over m and n.

We construct three independent vector solutions of Eq. (8.1) in the form

$$\boldsymbol{L} = \nabla\Psi\,, \quad \boldsymbol{M} = \nabla\times\boldsymbol{a}\,\Psi\,, \quad \boldsymbol{N} = \frac{1}{k}\nabla\times\boldsymbol{M} \tag{8.7}$$

where $\boldsymbol{a}$ is assumed to be a constant vector of unit length (generally, this vector need not be constant).

We assume that the vector potential $\boldsymbol{A}$ can be represented by an expansion in characteristic vector functions of the form

$$\boldsymbol{A} = \frac{\mathrm{i}}{\omega}\sum_n (a_n\boldsymbol{M}_n + b_n\boldsymbol{N}_n + c_n\boldsymbol{L}_n) \tag{8.8}$$

and the field vectors $\boldsymbol{E}$ and $\boldsymbol{H}$ may be similarly expressed by

$$\boldsymbol{E} = -\sum_n (a_n\boldsymbol{M}_n + b_n\boldsymbol{N}_n)\,, \quad \boldsymbol{H} = -\frac{k}{\mathrm{i}\omega\mu}\sum_n (a_n\boldsymbol{N}_n + b_n\boldsymbol{M}_n)\,. \tag{8.9}$$

In Eqs (8.8) and (8.9) a_n, b_n, c_n are the coefficients which have to be determined from the current distribution and $\boldsymbol{M}_n$, $\boldsymbol{N}_n$, $\boldsymbol{L}_n$ are components of the vector functions $\boldsymbol{M}$, $\boldsymbol{N}$, $\boldsymbol{L}$, respectively. The vector components $\boldsymbol{M}_n$, $\boldsymbol{N}_n$, $\boldsymbol{L}_n$ are associated with each characteristic scalar function Ψ_n which can be determined from Eq. (8.3).

First of all we will analyze the $\boldsymbol{M}$ component of the solution. We try to find a solution of Eq. (8.1) using

$$\boldsymbol{M} = \nabla\times[\boldsymbol{r}u(R)\Psi] = \nabla\Psi\times\boldsymbol{r}u(R) + \Psi\nabla\times\boldsymbol{r}u(R) = \boldsymbol{L}\times\boldsymbol{r}u(R) \tag{8.10}$$

where $\boldsymbol{r}$ is the unit vector in the direction of increasing R, $\boldsymbol{L}$ is given by Eq. (8.7), and $u = u(R)$ is yet an unknown scalar function of R. (Instead of a constant vector $\boldsymbol{a}$ in Eq. (8.7) we use the vector function $\boldsymbol{r}u(R)$ [146].) The components M_R, M_Θ, M_Φ of $\boldsymbol{M}$ in spherical coordinates R, Θ, Φ are

$$M_R = 0\,, \quad M_\Theta = \frac{1}{R\sin\Theta}\frac{\partial}{\partial\Phi}(u\Psi)\,, \quad M_\Phi = -\frac{1}{R}\frac{\partial}{\partial\Theta}(u\Psi)\,. \tag{8.11}$$

Therefore, the function $\boldsymbol{r}u(R)$ enable us to find the solution which is tangential to the sphere. The term $\boldsymbol{\nabla}^2\boldsymbol{C}$ on the left hand side of Eq. (8.1) has to be written as $\boldsymbol{\nabla\nabla.C} - \boldsymbol{\nabla}\times\boldsymbol{\nabla}\times\boldsymbol{C}$. The vectors $\boldsymbol{M}$ and $\boldsymbol{N}$ are solenoidal, i.e. their divergence $\boldsymbol{\nabla.M} = 0$ and $\boldsymbol{\nabla.N} = 0$. The wave equation (8.1) can be written in the form

$$\boldsymbol{\nabla}\times\boldsymbol{\nabla}\times\boldsymbol{M} - k^2\boldsymbol{M} = 0 \tag{8.12}$$

and expanded into R, Θ, and Φ components. Using Eq. (8.11) one gets in spherical coordinates

$$\boldsymbol{\nabla}\times\boldsymbol{\nabla}\times\boldsymbol{M} - k^2\boldsymbol{M} =$$

$$= \begin{vmatrix} \dfrac{\boldsymbol{r}}{R^2\sin\Theta}\,, & \dfrac{\boldsymbol{\vartheta}}{R\sin\Theta}\,, & \dfrac{\boldsymbol{\varphi}}{R} \\ \dfrac{\partial}{\partial R}\,, & \dfrac{\partial}{\partial\Theta}\,, & \dfrac{\partial}{\partial\Phi} \\ \dfrac{1}{R\sin\Theta}\left[\dfrac{\partial}{\partial\Theta}(\sin\Theta M_\Phi) - \dfrac{\partial}{\partial\Phi}(M_\Theta)\right], & -\dfrac{\partial}{\partial R}(RM_\Phi)\,, & \sin\Theta\dfrac{\partial}{\partial R}(RM_\Theta) \end{vmatrix} -$$

$$-k^2\boldsymbol{M} = 0 \tag{8.13}$$

where $\boldsymbol{\vartheta}$ and $\boldsymbol{\varphi}$ are the unit vectors connected with the coordinates Θ and Φ, respectively.

Using Eq. (8.11) we find out that the R component in Eq. (8.13) is identically zero

$$(\boldsymbol{\nabla}\times\boldsymbol{\nabla}\times\boldsymbol{M})_R = 0 \quad \text{and} \quad M_R = 0\,.$$

The tangential components in the direction of $\boldsymbol{\vartheta}$ and $\boldsymbol{\varphi}$ are subject to the same condition. This condition for the tangential components is satisfied if

$u(R)$ is such that

$$\frac{\partial^2}{\partial R^2}(u\,\Psi)+\frac{1}{R^2\sin\Theta}\frac{\partial}{\partial\Theta}\left[\sin\Theta\frac{\partial}{\partial\Theta}(u\,\Psi)\right]+\frac{1}{R^2\sin^2\Theta}\frac{\partial^2}{\partial\Phi^2}(u\,\Psi)+k^2u\,\Psi=0\,. \tag{8.14}$$

For $u(R)=R$ Eq. (8.14) yields the relation

$$\frac{1}{R^2}\frac{\partial}{\partial R}\left(R^2\frac{\partial\Psi}{\partial R}\right)+\frac{1}{R^2\sin\Theta}\frac{\partial}{\partial\Theta}\left(\sin\Theta\frac{\partial\Psi}{\partial\Theta}\right)+\frac{1}{R^2\sin^2\Theta}\frac{\partial^2\Psi}{\partial\Phi^2}+k^2\Psi$$
$$=\nabla^2\Psi+k^2\Psi=0\,. \tag{8.15}$$

It follows from the above that in spherical coordinates Eq. (8.12) is satisfied by

$$\boldsymbol{M}=\nabla\times\boldsymbol{R}\Psi=\boldsymbol{L}\times\boldsymbol{R}=\frac{1}{k}\nabla\times\boldsymbol{N} \tag{8.16}$$

where $\boldsymbol{R}=\boldsymbol{r}R$.

As follows from Eq. (8.11) for $u=R$ the components of $\boldsymbol{M}$ in spherical coordinates are given by

$$M_R=0\,,\quad M_\Theta=\frac{1}{\sin\Theta}\frac{\partial\Psi}{\partial\Phi}\,,\quad M_\Phi=-\frac{\partial\Psi}{\partial\Theta}\,. \tag{8.17}$$

From the relation

$$\boldsymbol{N}=\frac{1}{k}\nabla\times\boldsymbol{M}$$

in Eq. (8.7) the components of $\boldsymbol{N}$ can be derived. We obtain

$$kN_R=\frac{1}{R\sin\Theta}\left[\frac{\partial}{\partial\Theta}(\sin\Theta M_\Phi)-\frac{\partial}{\partial\Phi}M_\Theta\right]$$
$$=-\frac{1}{R^2\sin\Theta}\left[\frac{\partial}{\partial\Theta}\left(\sin\Theta\frac{\partial\Psi}{\partial\Theta}\right)+\frac{1}{\sin\Theta}\frac{\partial^2\Psi}{\partial\Phi^2}\right]R\,. \tag{8.18}$$

Using equation $\nabla^2\Psi+k^2\Psi=0$ in spherical coordinates one gets from Eq. (8.18)

$$kN_R=R\left[\frac{1}{R^2}\frac{\partial}{\partial R}\left(R^2\frac{\partial\Psi}{\partial R}\right)\right]+k^2R\Psi=\frac{\partial^2(R\Psi)}{\partial R^2}+k^2R\Psi\,. \tag{8.19}$$

After multiplication by R Eq. (8.19) can be written in the form

$$R^2\frac{\mathrm{d}^2\Psi}{\mathrm{d}R^2}+2R\frac{\mathrm{d}\Psi}{\mathrm{d}R}+\left(k^2R^2-p^2\right)\Psi=0 \tag{8.20}$$

where $p^2 = kRN_R/\Psi$. Eq. (8.20) can be satisfied by radial functions for $p = n(n+1)$, $n = 0, 1, 2, \ldots$. One may get for N_R

$$N_R = \frac{n(n+1)}{kR}\Psi\,. \tag{8.21}$$

The remaining two components of $\boldsymbol{N}$, i.e. N_Θ and N_Φ are given by

$$kN_\Theta = -\frac{1}{R}\frac{\partial}{\partial R}(RM_\Phi) = \frac{1}{R}\frac{\partial^2(R\Psi)}{\partial R\partial\Theta} \tag{8.22}$$

$$kN_\Phi = \frac{1}{R}\frac{\partial}{\partial R}(RM_\Theta) = \frac{1}{R\sin\Theta}\frac{\partial^2(R\Psi)}{\partial R\partial\Phi}\,. \tag{8.23}$$

We can write $\boldsymbol{M} = \boldsymbol{m}\exp(-\mathrm{i}\omega t)$ and $\boldsymbol{N} = \boldsymbol{n}\exp(-\mathrm{i}\omega t)$. Splitting off the time factor $\exp(-\mathrm{i}\omega t)$ we get from Eqs (8.3), (8.17), and (8.21)–(8.23) [146]

$$\begin{aligned}\boldsymbol{m}_{m,n\left\{\substack{\mathrm{e}\\ \mathrm{o}}\right\}} &= \mp\frac{m}{\sin\Theta}z_n(kR)P_n^m(\cos\Theta)\left\{\begin{array}{c}\sin(m\Phi)\\ \cos(m\Phi)\end{array}\right\}\boldsymbol{\vartheta}\\ &\quad - z_n(kR)\frac{\partial P_n^m(\cos\Theta)}{\partial\Theta}\left\{\begin{array}{c}\cos(m\Phi)\\ \sin(m\Phi)\end{array}\right\}\boldsymbol{\varphi}\end{aligned} \tag{8.24}$$

$$\begin{aligned}\boldsymbol{n}_{m,n\left\{\substack{\mathrm{e}\\ \mathrm{o}}\right\}} &= \frac{n(n+1)}{kR}z_n(kR)P_n^m(\cos\Theta)\left\{\begin{array}{c}\cos(m\Phi)\\ \sin(m\Phi)\end{array}\right\}\boldsymbol{r}\\ &\quad + \frac{1}{kR}\frac{\partial}{\partial R}\left[Rz_n(kR)\right]\frac{\partial P_n^m(\cos\Theta)}{\partial\Theta}\left\{\begin{array}{c}\cos(m\Phi)\\ \sin(m\Phi)\end{array}\right\}\boldsymbol{\vartheta}\\ &\quad \mp\frac{m}{kR\sin\Theta}\frac{\partial}{\partial R}\left[Rz_n(kR)\right]P_n^m(\cos\Theta)\left\{\begin{array}{c}\sin(m\Phi)\\ \cos(m\Phi)\end{array}\right\}\boldsymbol{\varphi}\,.\end{aligned} \tag{8.25}$$

A homogeneous continuous oscillating spherical dipole layer in a homogeneous medium has $n = 0$, $m = 0$, $P_0 = 1$. It follows from Eqs (8.24) and (8.25) that

$$\boldsymbol{m}_{m,n} = 0\,, \qquad \boldsymbol{n}_{m,n} = 0\,, \qquad \text{for } m = 0,\ n = 0 \tag{8.26}$$

i.e. such a layer does not generate a solenoidal electromagnetic field.

We examined a macroscopic spherical model with uniform continuous dipole layer. In this case $\boldsymbol{M} = 0$ and from Eq. (8.16) we determine that $\boldsymbol{L}$ can have only the radial component. The $\boldsymbol{L}$ vector is given by the gradient of the scalar function Ψ given by Eq. (8.3). As the positive and the negative charges of the dipole layer are equal and their distribution on the spherical surface is uniform no field of this type is generated.

A continuous homogeneous dipole layer is only an approximation valid for systems whose dimensions are much greater than the distance between the oscillating dipoles. The oscillating dipoles in cells are discrete sources, e.g. the protein molecules embedded in a cellular membrane. The electromagnetic field is generated only in the close vicinity of such an array of discrete sources. The smaller the distance between oscillating dipoles the smaller the energy of the electromagnetic field. A theoretical limit case for the distance decreasing to zero is the continuous dipole layer.

8.3 Computer Simulation of the Generated Electromagnetic Field

We assume the electromagnetic field to be generated by a large number of mutually coupled discrete sources – oscillating units forming oscillation centres. The oscillation centres in a cell are, therefore, analogous to phased array antena systems distributed along a closed surface. The cell generated electromagnetic field is an interference field generated by a large number of elementary generators. In the lateral direction the electromagnetic field generated by living cells has a mesoscopic tesseral structure corresponding to the distribution of the oscillation centres – oscillating molecules and structures. Between the oscillation centres the intensity of the electric field has the opposite direction with respect to the direction of the intensity along the axes of the oscillation centres. The distance between the oscillation centres may be of

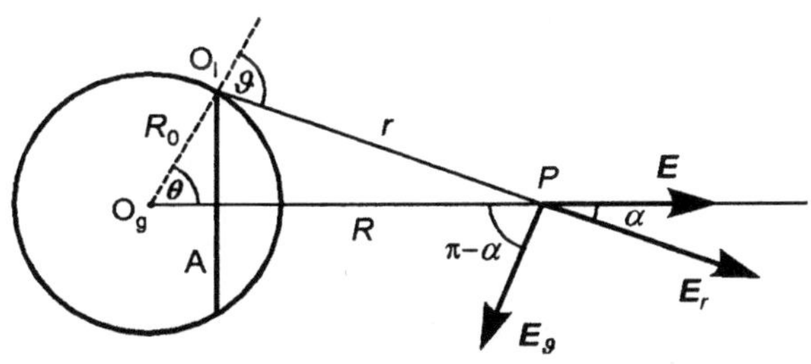

Fig. 8.3 A schematic picture of the geometry of a spherical model cell for evaluation of the intensity of the electric field. r, ϑ are the coordinates of the oscillation centres (the local coordinates). R, Θ are the coordinates of the model cell (the general coordinates). P is the point of observation. A is one of the plane cross-sections across the model cell. O_g and O_l are the centres of the general coordinates and of the local coordinates, respectively. A similar geometry is used for evaluation of the electric field of the ellipsoidal cells.

the order of 0.01–0.1 μm. We will determine the dependence of the intensity of the electric field on the distance along the normal to the surface of the model cell.

We will use two types of model cells for computer simulation of the generated electromagnetic field: the spherical models and the ellipsoidal models.

The electric components of the electromagnetic field generated by model cells are determined. The dipole field is computed using the well known relations for the electric dipole given by Eqs (13.63).

The oscillations of the discrete oscillation centres are assumed to be in phase. The oscillation centres are distributed quasi-uniformly on the surfaces of model cells. We make parallel cross-sections of a model cell. The cross-sections are perpendicular to the axis of rotation symmetry of the model cell (Fig. 8.3). The intersection curve of the cross-section plane with the surface of the model cell is a circle. The oscillation centres are positioned on the intersection curves. The axes of the oscillation centres of the dipoles are perpendicular to the cell surface. Uniform and continuous dipole distribution along the intersection curve is assumed. The total dipole moment is equal to the sum of the moments of the equidistant discrete dipoles. The distance between the parallel cross-section planes along the circumference is the same but not accurately equal to the given value for the cell. The continuous dipole distribution along the intersection circle in the spherical model cell has the rotation symmetry with respect to a line through the points of observation. This line (symmetry axis) intersects the centre of the spherical model and passes through two dipoles on the opposite sides of the surface of the model cell.

TABLE 2 Complex relative permittivity $\varepsilon' + i\varepsilon''$ of tissues with high water content in the frequency range from 10 MHz to 1000 MHz

f (MHz)	10	100	1000
ε'	81	81	81
ε''	900	130	23

TABLE 3 Complex relative permittivity $\varepsilon' + i\varepsilon''$ of water in the frequency range from 0.1 THz to 10 THz

f (THz)	0.1	1.0	10
ε'	8.5	4.5	2.0
ε''	12.5	2.5	1.0

In computation procedures we use the coordinates of the model cell (the general coordinates) and the coordinates of individual dipoles – oscillation centres (the local coordinates). The elementary transformations between the coordinate systems are carried out. The resulting intensity of the electric field in each point of observation is given by a sum of all vector contributions generated by the dipoles. We do not use any approximation. The accuracy of the results depends on the numerical accuracy of the computer.

8.4 Spherical Model Cells

We determine the electromagnetic field generated by a spherical cell with the uniformly distributed oscillation centres.

Distribution of the oscillation centres on the model cell surface is described in the previous section. The axes of the oscillation centres are in radial directions. We assume that in the frequency range 10–1000 MHz and 0.1–10 THz the properties of the ambient medium are the same as that of tissues with the high water content and of water, respectively. The complex relative permittivity of tissues with the high water content is given in Table 2 and the complex permittivity of water is in Table 3 [147]. (We are aware that properties of the cytosol and of the intercellular liquid medium depend on biological conditions and that their permittivities may differ from the values given in Tables 2 and 3.) The dipole moment of the oscillation centres is $e\ell$ where e is the elementary charge and $\ell = 5\,\text{nm}$. The radius of the model cell is $5\,\mu\text{m}$.

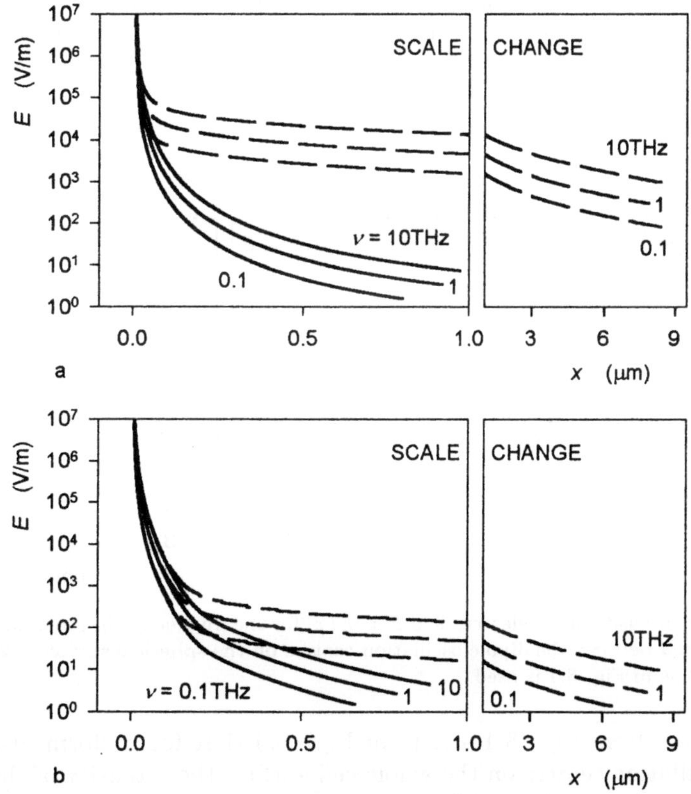

Fig. 8.4 The amplitude E of the intensity of the electric field in the radial direction as a function of the distance x from the spherical model cell surface. The frequency is a parameter (0.1, 1, 10 THz). The oscillation centres are distributed uniformly on the whole surface (the full curves) and only on a part of the spherical surface visible from the point of observation (the dashed curves). The distance d between the oscillation centres is 20 nm (a) and 200 nm (b).

Fig. 8.4 [148] shows the amplitude of the intensity of the electric field in the radial direction as a function of the distance from the spherical model cell surface. The parameter of the curves is frequency (0.1, 1, and 10 THz). The distance d between the oscillation centres is 20 nm (a) and 200 nm (b). The intensity of the electric field generated by the dipoles distributed uniformly on the whole surface (the full curves in Fig. 8.4) is considerably smaller than the intensity generated by the dipoles distributed only on a part of the model cell surface visible from the point of observation, i.e. on the spherical surface of a segment (the dashed curves in Figs 8.4 and 8.5).

Fig. 8.5 [148] shows the dependence $E = E(x)$ for frequency 10, 100, and 1000 MHz. The intensity of the electric field is smaller than that shown in Fig. 8.4a for model systems in the millimeter and in the submillimeter wavelength band. In the frequency range from 10 to 1000 MHz the permittivity of the medium is nearly two orders greater than in the submillimeter region (Table 2 and Table 3). Nevertheless, the energy losses in the submillimeter region may be considerably high on account of the absorption on discrete spectral lines.

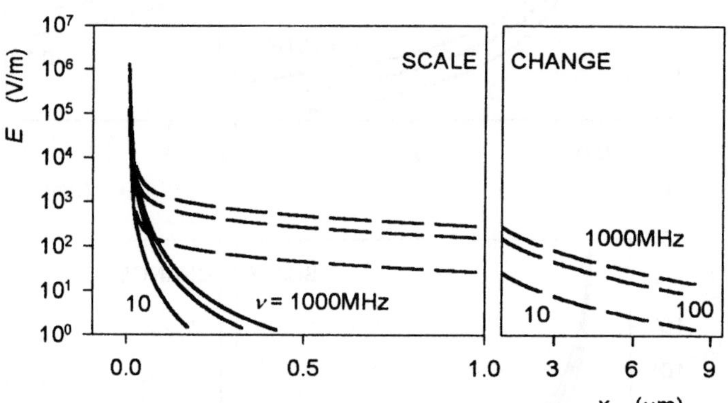

Fig. 8.5 The intensity of the electric field $E = E(x)$ for the frequency 10, 100, and 1000 MHz. The distance d between the dipole oscillation centres on the spherical surface is 20 nm. The same model as in Fig. 8.4 is used.

It follows from Fig. 8.4 and from Fig. 8.5 that for uniform distribution of the oscillation centres on the whole cell surface the intensity of the electric field has significant values within a distance of about 0.1 μm from the surface. The intensity steeply decreases with increasing distance from the surface of the model cell. If the oscillation centres are distributed only on a part of the surface the intensity of the generated electric field may be – in certain directions – much greater even at a distance of several μm. The decrease of the intensity with the distance is not so steep.

8.5 Ellipsoidal Model Cells

We determine the intensity of the generated electromagnetic field for uniform distribution of the oscillation centres on the surface of the model ellipsoidal cells. The amplitude of the intensity of the electric field as a function of the distance from the model cell surface is evaluated for rotational ellipsoides in the direction along the rotation axes (Fig. 8.6). Fig. 8.6a shows the $E = E(x)$ dependence for an elongated ellipsoide. Similar results for a disc ellipsoide are given in Fig. 8.6b. The disc ellipsoide has dimensions corresponding to the red blood cell (the semiaxes are 4 µm and 1.4 µm). The distance d between the oscillation centres is 200 nm and the dipole moment is $e\ell$ where $\ell = 5$ nm.

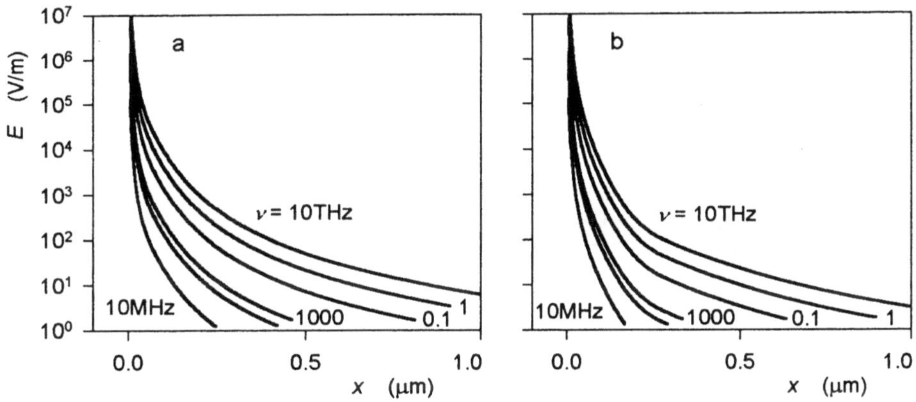

Fig. 8.6 The intensity of the electric filed $E = E(x)$ for an ellipsoide in the direction of rotation axis. The frequency ν is a parameter of the curves (10, 100, 1000 MHz and 0.1, 1, 10 THz). The distance between oscillation centres is 200 nm.
a) An elongated ellipsoide with semiaxes $a = 25$ µm, $b = 2.5$ µm.
b) A disc ellipsoide with semiaxes $a = 4$ µm, $b = 1.4$ µm.

The distribution of the oscillation centres is uniform (as is described in Section 8.3). The curves of the intensity of the electric field as a function of the distance for the spherical and for the ellipsoidal model cell (shown in Figs 8.4, 8.5, and 8.6) roughly coincide in the distances smaller than about 0.1 µm. In the close vicinity of the model cell surface the intensity of the electric field can be as high as 10^7 V/m. We can conclude that regardless of the shape of the cell the intensity of the electric field steeply decreases with the distance from the surface.

The actual distribution of the oscillation centres on the ellipsoidal surface may not be uniform to satisfy the condition of the minimum energy. We may expect that for disc shape ellipsoides the dipole density depends on the distance from the axis of the rotation symmetry, e.g. in the vicinity of the axis of the rotation symmetry the dipole density may be greater than in the more distant areas at the circumference. This non-uniform distribution can result from repulsion between oscillation centres positioned on the surface of the disc shape ellipsoide at its circular circumference. Intensity of the electric field at the symmetry axis (or near to it) may be greater than the intensity in other areas. This space distribution of the intensity of the electric field may be the cause of a special interaction of erythrocytes leading to rouleau formation (the coin like columns of the red blood cells). The shape and the dimensions of an erythrocyte correspond to the disc ellipsoidal model cell (except for the concave region of the erythrocyte around the disc rotation axis). Rowlands et al. measured interaction forces between erythrocytes [13–23,149] and assumed the long range interaction forces among erythrocytes resulting in rouleau formation to be caused by the Fröhlich electromagnetic interaction.

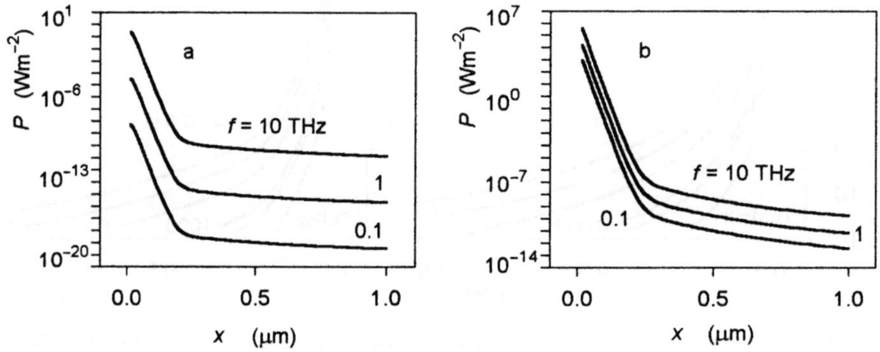

Fig. 8.7 Intensity of the energy flow in the region between oscillation centres as a function of the distance from the model cell surface. a) Evaluated for the $1/R$ term. b) Evaluated for all the dipole terms.

Interaction among cells may change the space distribution of the movable oscillation centres in membranes. The attractive interaction forces can increase the density of the oscillation centres in the regions between the cells which results in further increase of the interaction forces. The space distribution changes of the oscillation centres caused by the intercellular interaction may be important for behaviour of the cells.

8.6 Energy Flow from a Cell

We assumed that some polar proteins in the membrane form oscillation centres as is shown in Fig. 8.2. The dipole as well as the multipole oscillation centres generate electromagnetic fields containing radiation component flowing outwards from the source. We will analyze the electric dipole oscillation centres and assess the energy flow from a cell. The main contribution to the energy flow in the direction perpendicular to the cell surface may be found in the region between oscillation centres. If the distance between two neighbouring oscillation centres is d then the greatest energy flow is of about $d/4$ apart from the axis of an oscillation centre. The mean intensity of the energy flow from a source is given by the real part P of the complex Poynting vector given by Eq. (13.64) (i.e. the energy per second crossing a unit area whose normal is oriented in the direction of the Poynting vector). The complex Poynting vector enable us to to assess the radiation component of the generated electromagnetic field.

We used a spherical model cell with a radius of 5 μm. The distance between oscillation centres $d = 0.1\,\mu\text{m}$, and the dipole moment $p = e\ell$ (where $\ell = 10\,\text{nm}$). The intensities of the electric field and of the magnetic field are computed as the vector sums of the contributions from all the oscillation centres on the cell surface. Figs 8.7a, b show P as a function of the distance from the model cell surface in the region $d/4$ apart from the axis of an oscillation centre. Fig. 8.7a contains radiation component evaluated only for the $1/R$ terms of the electric and magnetic intensities given by Eqs (13.63). Fig. 8.7b shows P evaluated for all the terms given by Eqs (13.63). Significant values of the energy flow are only in the vicinity of the model cell surface. We note that P is determined only for a small region in the distance $d/4$ from the axis of the oscillation centre. The area of this region is smaller then $\pi(d/4)^2 \approx 10^{-15}\,\text{m}^2$. If the intensity of the power flow $P = 10^{-12}\,\text{Wm}^{-2}$ the corresponding power crossing the area is smaller than $10^{-27}\,\text{W}$. Energy of a photon at $10^{13}\,\text{Hz}$ is $h\nu = 6.6 \times 10^{-21}\,\text{J}$. Therefore, in the region around the oscillation centre one photon is emitted within an average time interval greater than about 70 days. The whole cell emits one photon (in arbitrary direction) within the time period greater than 10 s. Nevertheless, the state with minimum emission of energy need not exist in all phases of the cell cycle. In the mitotic phase the cell can radiate a strong electromagnetic field.

8.7 Electric Field Generated by Microtubules

The microtubules are bound to the proteins in the plasma membrane and form a common functional system with them. Therefore, vibrations in the membrane proteins are connected with vibrations in the structures of the cytoskeleton. Analysis of the dynamic processes in the microtubules and of the generated electric field may help us to understand the cellular electric activity.

The tubulin heterodimers are polar and the vibrations create oscillating dipoles or oscillating multipoles. The polar vibrations generate an electric field. We will adopt a linear model of a microtubule with a length of $L_m = 3\,\mu m$ where each heterodimer carries an oscillating electric dipole $p = be\ell_h$ (e is the elementary charge, $\ell_h = 8\,nm$, and $b = b(x)$ where x is the distance along the microtubule axis). Except for b the dipole moments are identical. The frequency of vibrations $\nu = 100\,MHz$ and the relative permittivity of the ambient medium $\varepsilon' = 81$. We evaluate the intensity of the electric field along the lines parallel with the microtubule axis.

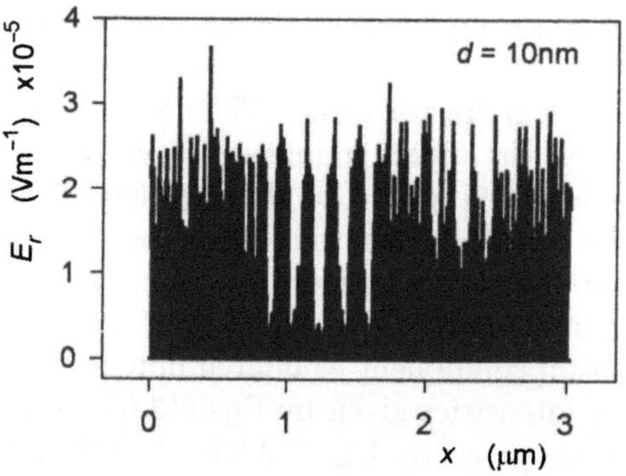

Fig. 8.8 Thermal vibrations. The absolute value of the intensity of the electric field E_r as a function of x.

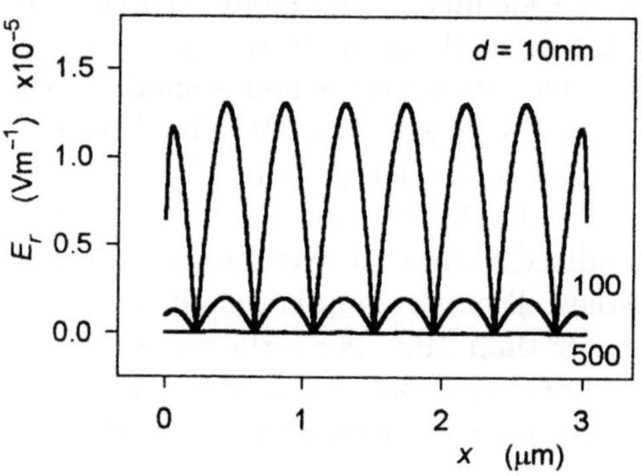

Fig. 8.9 Coherent vibrations of a vibration mode. The absolute value of the intensity of the electric field E_r as a function of x.

A typical feature of the vibrations in thermodynamic equilibrium is their randomness. Using classical concepts we may state that the "amplitude" and the "phase" have no coherent components. The intensity of the electric field for random orientations of the dipole moments is given in Fig. 8.8. E_r shown in Fig. 8.8 represents the absolute value of the intensity in the radial direction perpendicular to the microtubule axis (the r direction). The distance d from the "surface" of the microtubule model is 10 nm. The greatest absolute value of the intensity is of the order of $10^5\,V/m$. At a distance 0.1 and 0.5 μm the peak values are about 10^2 and $1\,V/m$, respectively. We used a generator of random numbers to determine the orientations of the dipole moments of particular heterdimers ($b = \pm 1$). The intensity of the electric field has random fluctuations in space and time and can decrease to very small values. In the random case the space-time distribution of the dipole moments should correspond to the maximum value of entropy (on account of the Second Law of Thermodynamics).

In thermodynamic equilibrium Planck's Law determines the distribution of energy in the normal modes. A vibration mode is equivalent to the harmonic oscillator which is subject to random disturbances. The resulting vibrations are given by superposition of the vibrations of all the modes. Nevertheless, under certain conditions the excited vibrations can be far from thermodynamic equilibrium with an important coherent component. In the coherent state certain mode (or modes) are excited; its (their) energy is far from the thermodynamic equilibrium value. An example is shown in Fig. 8.9. Fig. 8.9 shows E_r of the vibration mode $k = 7\pi/L_m$ as a function of x. The distance d is 10 nm, 0.1, and 0.5 μm. The vibrations generate a dynamic periodic potential around the microtubule. The k value may be considerably greater than that used in Fig. 8.9 and the period of the standing waves may be comparable in size to the heterodimers. The intensity of the electric field depends on the excitation of vibrations in the microtubule.

9. Information Transfer Between Oscillation Systems

9.1 Interaction and Information

Transfer of energy and mass in biological systems enables interaction and communication between different entities. Interaction is an important tool for cooperation and creation of large organized systems composed of many cooperating subsystems. Transfer of energy may have many forms and may include the transfer of electrostatic energy, energy of low frequency oscillating fields, energy of light, energy of vibrations, etc. Energy may be transferred by means of molecules having a high energy content; the molecule is transported to the target entity where its energy is released. Molecules can contain energy in the chemical bonds, in the excited electron states, in the conformation states, etc. In biological systems we can find a wide diversity of the mass transport in which the mass by itself carries information. The transport of the phosphate groups is a special case of the utmost importance. Phospory-lation of proteins has fundamental effects on the biological activity. It is difficult to describe the variety of the interaction mechanisms participating in organization. A common measure of the interaction leading to cooperative behaviour is the transfer of information. We will analyze the transfer of information between polar coherent vibration systems. As a model the Fröhlich coherent vibration systems will be used.

The Fröhlich coherent vibrations are polar and generate a coherent electromagnetic field. We will not analyze here the effects of ions, charged particles, and polar molecules in the surroundings of the Fröhlich system on the generated electromagnetic field. We remark that the generated electromagnetic field is not screened by ions and moving charged particles as the frequency of the vibrations is high enough with respect to the density and mobility of ions in the cytosol and in the intercellular liquid medium. (Screening depends on the relaxation time $\tau = \varepsilon/\sigma$ where ε is the permittivity and σ is the conductivity; Appendix 13.13.) The same holds for the biological molecules. (The fundamental gap in the energy band structure of proteins is of about 8 eV or higher. On account of the electron hopping mechanism proteins are not good isolators; their conductivity is not negligible and may reach the value of 10^{-3}–$10^{-2}\,\Omega^{-1}\mathrm{cm}^{-1}$ [150]. Nevertheless, the hopping conductivity is not high enough for effective screening in the frequency bands in question.) The electric component of the electromagnetic field polarizes the ambient medium and acts by force on electric charges. The mediatory role of the electromagnetic field is connected with the transfer of information. The information transmission forms a background in cooperative phenomena and in creation of the space-time coherent regions. The information transmission mediated by the Fröhlich electromagnetic field can play a fundamental role in the selforganization of biological systems. As the information transfer is connected with an electromagnetic field the information theory developed for the communication purposes may be used [151–155]. We have to mention that any physical system of information transfer has an upper limit of the number of the information functions [156], i.e. it can transfer a finite amount of information. In this case finitely many outcomes of independent events are assumed.

The spectral components of the electromagnetic field correspond to the frequencies of normal modes. The stationary nature of the frequency and of the energy of a normal mode is an important property (the excitation of a harmonic oscillator with time dependence of the type $\sin(\omega t + \varphi)$ for a limited time period does not represent excitation of a single normal mode). We will assume here that the normal modes of the Fröhlich system satisfy this condition of stationarity. From a purely theoretical point of view the Fröhlich system is not a finite system. The energy of the normal modes need not acquire the values in a limited range, i.e. there are not finitely many outcomes. (Nevertheless, in real oscillation systems the energy has to be limited; if the energy exceeds certain upper limit the molecules change their structure, dissociate, etc.) Energy distribution is given by the Bose–Einstein statistics $[B \exp(\beta\nu) - 1]^{-1}$ where B is a constant. The probabilities of the different energy levels of a normal mode are not equal and may differ in various normal modes. The noise power in the Fröhlich system as well as in a communication system determines the energy levels that may be distinguished. The information flow between the Fröhlich coherent vibration systems depends on the number of the normal modes, on their energy and frequency, and on the surrounding medium between the Fröhlich systems (i.e. on the communication channel).

The Fröhlich coherent states are very likely excited in the protein molecules in the cellular membranes and in the cytoskeleton binding these proteins as is suggested by the experimental results; a short survey may be found e.g. in [129,157]. Proteins are polar and some of them use the metabolic energy, which may excite coherent states. The electromagnetic field generated by one oscillation centre is "received" by other oscillation centres and influences their oscillations. Information is, therefore, transferred between the oscillation centres. Information transferred between the oscillation centres depends on the amount of information in the "transmitter" oscillation centre and on the information loss in the channel (i.e in the medium between the oscillation centres). The information loss in the channel between two Fröhlich systems is caused by spreading and attenuation of the electromagnetic wave and by increase of noise. If the information transfer is great enough the oscillation centres in a cell can form a phased array system (e.g. a system of harmonic oscillators whose vibrations maintain certain phase relations). The information transfer may exist among the oscillation centres in different cells too. Information can be transferred across a channel between neighbouring cells. The channel filled with the intercellular liquid may be e.g. about 15 nm wide.

The information transfer mediated by the Fröhlich electromagnetic field in biological systems has some extraordinary properties in comparison with other means and mechanisms of the information transfer, e.g. in comparison with chemical mechanisms. The transfer of ions and molecules is very slow. The transfer by an electromagnetic wave is extremely fast; the time delay for information transfer across a channel of 15 nm wide between two cells is of the order of 10^{-17} s. Nevertheless, we have to remark that information may be transferred through the cytoskeleton components (i.e. through the microtubules, the actin filaments, and the intermediate filaments) inside a cell, and through the connecting fibres and through the extracellular matrix between cells. These structures may form communication channels too. This information transfer can be also of the electric nature but the time delay is very likely several orders greater than that for the transfer by an electromagnetic wave.

Due to the resonance effects selectivity of the information transfer may occur. The information transfer mediated by the Fröhlich electromagnetic field depends on the surrounding medium. If the frequencies of certain spectral lines are e.g. in the region of anomalous dispersion with the strong attenuation the information transfer is cut off. External electromagnetic fields might have certain effects on the information transfer too.

9.2 Energy Condensation, Probability and Information

Energy condensed in a particular mode of the Fröhlich coherent system depends on the energy supply (i.e. on S) and on the linear and on the non-linear transition probabilities (i.e. on Φ and on χ) – Eq. (5.11). The probability density of various energy levels is a function of the probability densities of S, of Φ, and of χ. This is the "inner" probability density which depends on the internal non-linear mechanism of the Fröhlich coherent vibration system.

Metabolic energy supply to a Fröhlich system has certain probability distribution. We may assume here e.g. that the distribution is normal with the average value much greater than the average value of the noise energy and the standard deviation much smaller than the average value. The probability density of various energy levels (a discontinuous type of the probability density) is a function of energy supply (of the coefficient of energy supply S). Using the function of the random variable $E_i = E_i(S)$ [136] (where E_i is the energy in the normal mode with the frequency ν_i) the probability density of energy in a normal mode may be derived. As the first order approximation we may assume that $E_i = aS + b$. In this case the probability density of E_i has the Gaussian form too. The energy supply in the region of the average value has a great probability density which may be close to 1 for a narrow Gaussian curve. The amount of information (determined from this probability density) is, therefore, very small. Nevertheless, the system generates a large electromagnetic field, has a great effect on other Fröhlich systems, and transfers a large amount of information. It follows from this controversy that the prob-

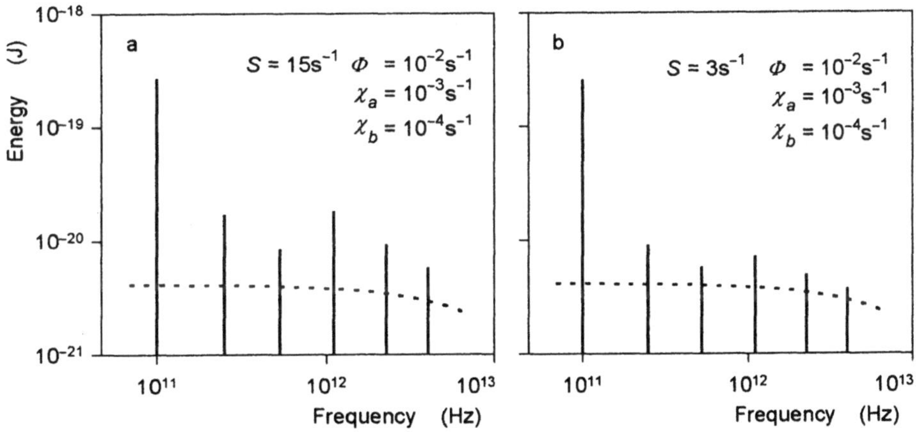

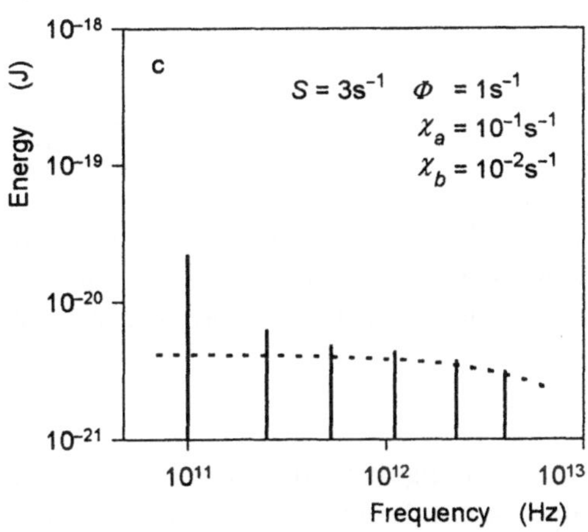

Fig. 9.1 Energy in the normal modes of a model Fröhlich system as a function of frequency. The horizontal dotted lines indicate the thermodynamic equilibrium energy.

a) The weak coupling to the heat bath ($\chi_a = 0.001\,\mathrm{s}^{-1}$, $\Phi = 0.01\,\mathrm{s}^{-1}$) and the great energy supply ($S = 15\,\mathrm{s}^{-1}$).

b) The weak coupling to the heat bath ($\chi_a = 0.001\,\mathrm{s}^{-1}$, $\Phi = 0.01\,\mathrm{s}^{-1}$) and the small energy supply ($S = 3\,\mathrm{s}^{-1}$).

c) The strong coupling to the heat bath ($\chi_a = 0.1\,\mathrm{s}^{-1}$, $\Phi = 1.0\,\mathrm{s}^{-1}$) and the small energy supply ($S = 3\,\mathrm{s}^{-1}$).

ability based on the internal mechanism of energy condensation cannot be used for information purposes.

Comparison of the Fröhlich system with a non-Fröhlich system in the surroundings (which is in thermodynamic equilibrium) may lead to more appropriate results. For instance, phase locking of two interacting oscillators depends on the energy transfer between them. The greater the energy over the thermodynamic equilibrium value the greater the energy transfer between the oscillators. The transfer of phase is the transfer of information: the energy transfer between the oscillators mediates the information transfer. The information transfer among the Fröhlich systems leads to creation of coherent regions, e.g. to harmonic oscillations in phase in certain areas.

Observing the Fröhlich system embeded in a non-Fröhlich medium we find out that certain normal modes are far from thermodynamic equilibrium while the remaining parts are in (or near to) thermodynamic equilibrium. Due to energy condensation occupation of the modes of the Fröhlich system has a deterministic and a random component. The deterministic component may be considerably greater than the random component and in this case we will assume that certain energy level is occupied in a deterministic way. But the system as a whole very likely remains in thermodynamic equilibrium. Assuming that the Fröhlich system is replaced by the non-Fröhlich system with the same energy distribution we may assign the thermodynamic probability to the energy levels of particular normal modes. This probability may be related to information. As a zero order approximation we will assume that the normal modes have definite discrete values of energy (as it is in the normal modes with condensed energy if the noise fluctuations are neglected). In this case the probability densities of the discrete energy levels may be given by the Boltzmann distribution (as in the classical approach). The probability density for a normal mode with the frequency ν_i may be written in the form

$$P_i = \frac{\exp\left(-\frac{E_i}{kT}\right)}{\sum\limits_{n_m=0}^{\infty} \exp\left(-\frac{n_m h\nu_i}{kT}\right)} \tag{9.1}$$

where $E_i = n_i h\nu_i$ is the energy and n_i, n_m are the occupation numbers. With respect to the definition of information the amount of information in a normal mode with frequency ν_i may be given by the relation

$$I_i = \lg \frac{1}{P_i} \, . \tag{9.2}$$

Assuming that the normal modes are mutually independent the total amount of information in the whole system is given by the sum of information in particular modes

$$I = \sum_i I_i = -\sum_i \lg P_i \tag{9.3}$$

where P_i is given by Eq. (9.1). (The assumption that normal modes in the Fröhlich systems with condensed energy are mutually independent is obviously not satisfied. Nevertheless, we will use it for the first rough assessment of the amount of information.)

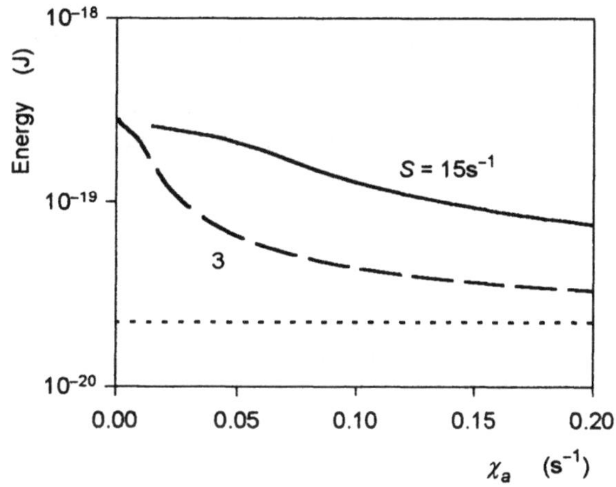

Fig. 9.2 The total energy in a model Fröhlich system as a function of the non-linear transition probability χ_a. The coefficient of energy supply S is a parameter. The horizontal dotted line indicates the thermodynamic equilibrium energy.

First of all we will compare the energy in vibration modes with their information content. We use model systems with six modes. The Φ_i and χ_{ij} transition probabilities are not uniform. The model vibration system has two groups of the normal modes. The low frequency group consists of three normal modes with frequencies ν_1, ν_2, ν_3; the high frequency group includes three remaining normal modes. The non-linear energy transfer (i.e. the transition probability χ_a) within these two groups of normal modes is greater than the energy transfer (i.e. the transition probability χ_b) between the groups ($\chi_b = 0.1\,\chi_a$). This model may correspond to a real system; e.g. the energy transfer between a normal mode in the millimetre wave band and a normal mode in the infrared band may be smaller than within these particular bands as was discussed in Chapter 5.

Fig. 9.1 shows the energy condensed in the model Fröhlich coherent vibration systems. The vibration system in Fig. 9.1a has the weak coupling to the heat bath ($\chi_a = 0.001\,\mathrm{s}^{-1}$, $\Phi = 0.01\,\mathrm{s}^{-1}$) and the great energy supply

($S = 15\,\mathrm{s}^{-1}$). The system is highly excited. The energy is condensed in the lowest frequency mode of the low frequency group of modes and in the quasi-lowest frequency mode (the lowest frequency mode of the high frequency group of modes). Fig. 9.1b shows the energy condensed in the same model system as Fig. 9.1a but with the small energy supply ($S = 3\,\mathrm{s}^{-1}$). The energy is condensed only in the lowest frequency mode. Fig. 9.1c shows the energy condensed in a model vibration system with the strong coupling to the heat bath ($\chi_a = 0.1\,\mathrm{s}^{-1}$, $\Phi = 1\,\mathrm{s}^{-1}$) and the small energy supply ($S = 3\,\mathrm{s}^{-1}$). Even the energy condensed in the lowest frequency mode is small. The strong non-linear coupling to the heat bath reduces energy condensation in the Fröhlich vibration systems.

Fig. 9.2 shows the total energy in the Fröhlich vibration system as a function of the non-linear transition probability χ_a. The total energy condensed in the Fröhlich systems strongly decreases with increasing values of the trans-

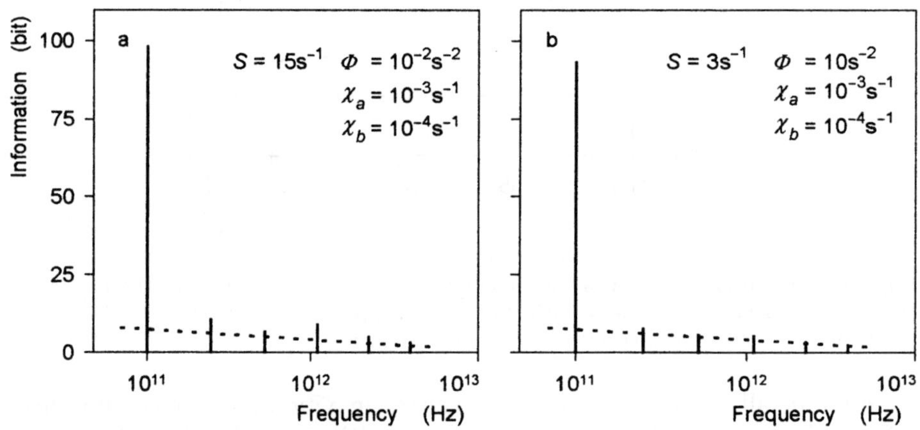

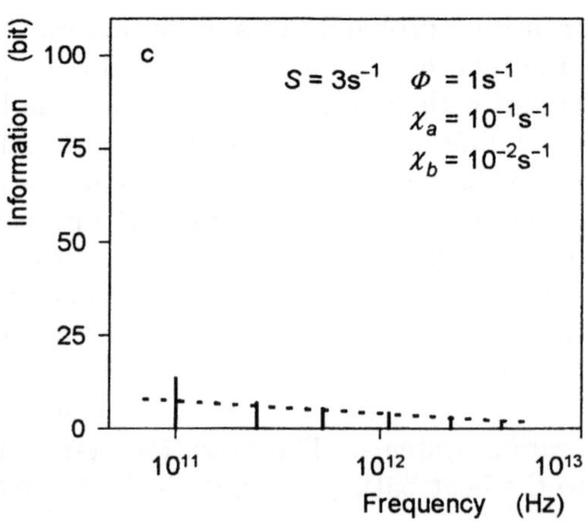

Fig. 9.3 Information content in the normal modes of a model Fröhlich system as a function of frequency. The Boltzmann probability is used. The dotted curves indicate the thermodynamic equilibrium values.

a) The weak coupling to the heat bath ($\chi_a = 0.001\,\mathrm{s}^{-1}$, $\Phi = 0.01\,\mathrm{s}^{-1}$) and the great energy supply ($S = 15\,\mathrm{s}^{-1}$).

b) The weak coupling to the heat bath ($\chi_a = 0.001\,\mathrm{s}^{-1}$, $\Phi = 0.01\,\mathrm{s}^{-1}$) and the small energy supply ($S = 3\,\mathrm{s}^{-1}$).

c) The strong coupling to the heat bath ($\chi_a = 0.1\,\mathrm{s}^{-1}$, $\Phi = 1\,\mathrm{s}^{-1}$) and the small energy supply ($S = 3\,\mathrm{s}^{-1}$).

ition probability χ_a which depends on the coupling of the system to the heat bath.

We used models with 6 normal modes. The energy condensed in the first mode and in the third mode may be greater than energy in the system in thermodynamic equilibrium. In real systems (e.g. in the protein molecules and structures) the number of normal modes is much greater as it depends on the number of atoms. The number of atoms in a protein may be of the order of 10^3. The total phonon energy may be, therefore, much greater than the energy condensed in some normal modes. The system temperature of the excited Fröhlich system need not differ considerably from the temperature before excitation, i.e. the majority of the vibration modes remains in or near thermodynamic equilibrium.

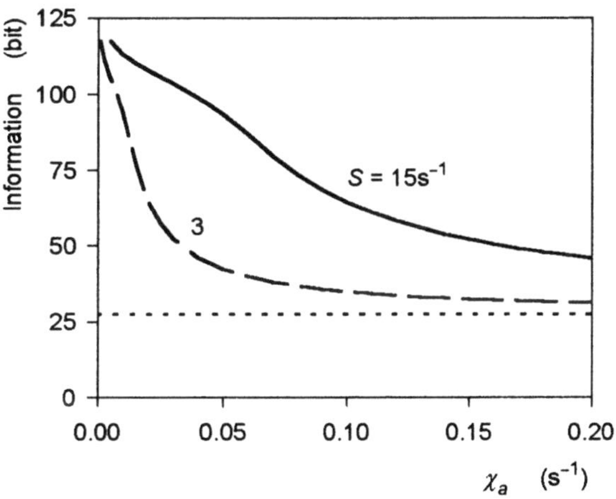

Fig. 9.4 Information content in a model Fröhlich coherent vibration system as a function of χ_a (i.e. as a function of the non-linear coupling to the heat bath). The dotted line indicates the thermodynamic equilibrium value.

Information content in the models of the Fröhlich coherent vibration systems was evaluated from Eq. (9.2). Fig. 9.3 shows information in particular normal modes for the same models as in Fig. 9.1. The Fröhlich system with the weak coupling to the heat bath and the great energy supply has a great information content in the lowest frequency mode and an enhanced information content in the quasi-lowest frequency mode (Fig. 9.3a). The same system with the small energy supply has a great information content only in the lowest frequency mode (Fig. 9.3b). The Fröhlich system with the strong coupling to the heat bath and with the small energy supply has a small information content even in the lowest frequency mode (Fig. 9.3c).

Information of the whole Fröhlich system as a function of the non-linear transition probability χ_a (i.e. of the non-linear coupling to the heat bath) is plotted in Fig. 9.4. The greater the coupling to the heat bath the smaller the information content of the Fröhlich system. For the great energy supply the decrease of information occurs at greater values of χ_a than for the small energy supply. For the strong coupling to the heat bath information content tends to the thermodynamic equilibrium value. The information content given above was evaluated under the assumption that the energy levels whose distance is $h\nu_i$ may be distinguished, i.e. that there is no noise in the system. A single energy quantum is much smaller than the average noise energy. Information should be evaluated for a system with the noise.

9.3 Information Entropy in Coherent Systems

A measure of information is connected with random events. The ensemble average, i.e. the expectation value of the information in the system

$$\langle I \rangle = -\sum_i P_i \lg P_i \tag{9.4}$$

is called the information entropy [37,42,137,158]. From the communication theory point of view the information entropy is a measure of uncertainty.

The coherent signal and the thermal noise are present in the Fröhlich system. The information entropy depends on the energy condensed in the Fröhlich system and on the noise energy. The information entropy is a function of all possible combinations of energy levels whose differences depend on the noise energy. One can distinguish energy levels whose difference is equal to or greater than the average noise energy. As the first order approximation the energy levels may be assumed to be equidistant and the difference between them given by the noise energy (for white noise). Nevertheless, the variance of the fluctuations of the energy level in the presence of noise may depend on the energy level too. Therefore, the energy levels are not equidistant.

The energy E_i of a normal mode with frequency ν_i is given by [37]

$$E_i = E_{\mathrm{s}i} + E_{\mathrm{n}i} + E_{\mathrm{sn}i} \tag{9.5}$$

where $E_{\mathrm{s}i}, E_{\mathrm{n}i}$, and $E_{\mathrm{sn}i}$ are the energy of the coherent signal, of the noise, and the energy generated by the interference of the coherent signal and of the noise, respectively. The energy of the coherent signal of the normal mode with the frequency ν_i is subject to the fluctuations whose mean square value is given by [37]

$$\Delta_i^2 = \langle (E_i - \langle E_i \rangle)^2 \rangle = h\nu_i \langle E_i \rangle + 2\langle E_i \rangle \langle E_{\mathrm{n}i} \rangle - \langle E_{\mathrm{n}i} \rangle^2 \tag{9.6}$$

where $\langle E_i \rangle, \langle E_{\mathrm{n}i} \rangle$ are the average values of the total energy and of the noise energy, respectively. $\langle E_i \rangle$ and $\langle E_{\mathrm{n}i} \rangle$ are given by the relations [37]

$$\langle E_i \rangle = E_{si} + \langle E_{ni} \rangle = \langle n_i \rangle h\nu_i \ , \qquad \langle E_{ni} \rangle = \frac{h\nu_i}{\exp\left(\frac{h\nu_i}{kT}\right) - 1} \ . \tag{9.7}$$

In [37] the energy levels of coherent signals are equiprobable (when entropy has the maximal value). Nevertheless, probability of a discrete particular energy level is given by the Boltzmann distribution. We will assign the Boltzmann distribution to the energy levels of coherent signals.

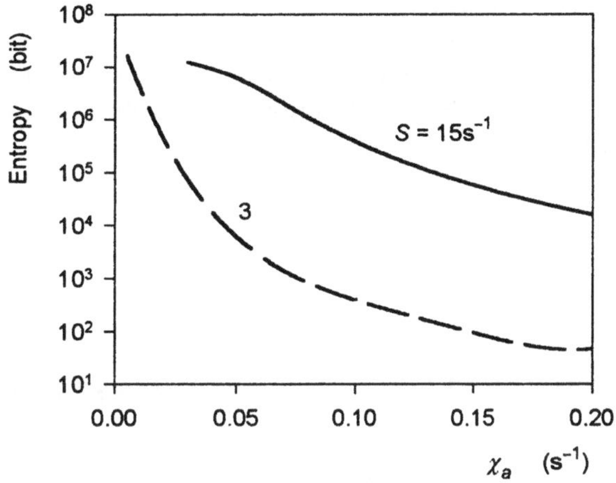

Fig. 9.5 Information entropy of a model Fröhlich system as a function of χ_a (i.e. as a function of the non-linear coupling to the heat bath).

The probability density of the energy level $\langle E_i \rangle$ may be written in the form

$$P_i = \frac{\sum\limits_{\langle E_i \rangle}^{\langle E_i \rangle + \Delta_i} \exp\left(-\frac{n_i h\nu_i}{kT}\right)}{\sum\limits_{n_m=0}^{\infty} \exp\left(-\frac{n_m h\nu_i}{kT}\right)} = \exp\left(-\frac{n_i h\nu_i}{kT}\right)\left\{1 - \exp\left[-\frac{(\Delta n_i + 1)h\nu_i}{kT}\right]\right\} \tag{9.8}$$

where $\Delta n_i = \Delta_i/(h\nu_i)$ and Δ_i is given by Eq. (9.6). (We assume here that Δ_i is an integer multiple of $h\nu_i$.)

The information entropy of a normal mode with the frequency ν_i may be given by

$$\langle I_i \rangle = \sum_{n_i=n_{i0}}^{n_{i\max}} \exp\left(-\frac{n_i h\nu_i}{kT}\right)\left[1 - \exp\left(-\frac{(\Delta n_i + 1)h\nu_i}{kT}\right)\right] \times$$

$$\times \left\{ \frac{n_i h\nu_i}{kT} \frac{1}{\ln 2} - \lg \left[1 - \exp\left(-\frac{(\Delta n_i + 1) h\nu_i}{kT} \right) \right] \right\} \tag{9.9}$$

where $n_{i\max} = E_{\max}/(h\nu_i)$ and $E_{\max}$ is the maximum value of the total energy.

In statistical physics the $E_{\max}$ value is not limited ($E_{\max} \to \infty$) and in this case Eq. (9.9) yields the correct value of the entropy. We confined our analysis to the finite values of $E_{\max}$. The $\langle I_i \rangle$ value may be considered as the greatest average information content of the Fröhlich coherent vibration system.

The Fröhlich coherent system has several modes with different frequencies and with various energy levels. Energy may be distributed in the normal modes arbitrarily. The total information entropy is given by

$$\langle I \rangle = \sum_{\langle E_i \rangle = \langle E_{ni} \rangle}^{E_{\max}} \left\{ P_i \lg \frac{1}{P_i} + \sum_{\langle E_j \rangle = \langle E_{nj} \rangle}^{E_{\max} - \langle E_i \rangle} \left[P_j \lg \frac{1}{P_j} + \cdots \right. \right.$$
$$\left. \left. \cdots + \sum_{\langle E_k \rangle = \langle E_{nk} \rangle}^{E_{\max} - \langle E_i \rangle - \langle E_j \rangle - \cdots} \left(P_k \lg \frac{1}{P_k} \right) \right] \right\} \tag{9.10}$$

where $P_i, P_j, \ldots P_k$ are given by Eq. (9.8) and the indexes i, $j, \ldots,$ k denote different normal modes.

Fig. 9.5 shows the information entropy $\langle I \rangle$ of the model Fröhlich system with six normal modes as a function of the non-linear transition probability χ_a. The information entropy is evaluated from Eq. (9.10). The model system has the same parameters as the system used in Figs 9.2 and 9.4. The information entropy depends on the energy condensed in the Fröhlich coherent system. The information entropy decreases with the increasing coupling to the heat bath (i.e. with increasing values of χ_a). The information entropy curve for the great energy supply ($S = 15\,\mathrm{s}^{-1}$) is similar to the curve for the small energy supply ($S = 3\,\mathrm{s}^{-1}$) but it is shifted to the greater values of χ_a.

9.4 Communication Channel Between Coherent Systems

The Fröhlich systems generate an electromagnetic field. The electromagnetic field transfers information through the medium between the Fröhlich systems and inside the Fröhlich system between the oscillating units. The medium in between represents a communication channel. The transmission of information takes place with noise disturbances originating in the channel medium.

We try to find a measure of the amount of information transmission through the channel between the Fröhlich systems or between oscillating units. We will use the concept of the information capacity of the channel [152].

To derive a relation for the information capacity we assume discrete input and discrete output ensembles applied to the communication channel, i.e. the sets of the a_m and b_j events of the ensembles A and B, respectively. The average mutual information is defined in the form [137,158,159]

$$I(A;B) = \sum_m \sum_j P(a_m, b_j) \lg \frac{P(a_m/b_j)}{P(a_m)} \tag{9.11}$$

where $P(a_m, b_j) = P(b_j)P(a_m/b_j)$, $P(a_m/b_j)$ is the conditional probability, and $P(a_m)$ and $P(b_j)$ are the probabilities of the event a_m and of the event b_j, respectively. The term $\lg[P(a_m/b_j)/P(a_m)]$ in Eq. (9.11) is the mutual information, i.e. the amount of information transferred between the event a_m and the event b_j. The mutual information may be used as a measure of the information decrease caused by the channel. If the conditional probability is smaller than one the mutual information is smaller than the input selfinformation $I(a_m) = -\lg P(a_m)$.

The average mutual information may be expressed as (Appendix 13.8 [158])

$$I(A;B) = \langle I(A) \rangle - \langle I(A/B) \rangle = \langle I(B) \rangle - \langle I(B/A) \rangle \tag{9.12}$$

where

$$\langle I(A) \rangle = -\sum_m P(a_m) \lg P(a_m) \tag{9.13}$$

is the information entropy (sometimes denoted as the average selfinformation) of the input ensemble and

$$\langle I(A/B) \rangle = -\sum_m \sum_j P(a_m, b_j) \lg P(a_m/b_j) \tag{9.14}$$

is the conditional information entropy of A ensemble given B ensemble. Similar relations hold for $\langle I(B) \rangle$ and $\langle I(B/A) \rangle$. $\langle I(B/A) \rangle$ is the average information loss specifying the noise disturbance in the channel.

Under the maximization condition of the $I(A;B)$ the capacity of the channel is defined as [158]

$$C_i = I(A;B) \qquad \text{(bits/event)}. \tag{9.15}$$

The capacity of the channel may be determined using the relation between the physical entropy and the information entropy. The information entropy means uncertainty. The thermodynamic entropy is a measure of disorder and its increment may be expressed as (Appendix 13.9 [37,158])

$$\Delta S = k \ln 2 \langle I(B/A) \rangle \, . \tag{9.16}$$

We use this equation to evaluate the average mutual information. The power spectral density of the noise is given by the Eq. (9.7) for $\langle E_{\mathrm{n}i} \rangle$. The maximum value of the entropy $\langle I(B) \rangle$ is reached when the microstates of the total emission correspond to the thermodynamic equilibrium distribution. Therefore, we assume that the signal energy is given by a relation similar to Eq. (9.7) but with temperature T_{si}. The output energy is given by the relation

$$\langle E_{\mathrm{o}i} \rangle = \frac{h\nu_i}{\exp\left(\dfrac{h\nu_i}{kT_{\mathrm{s}i}}\right) - 1} = \langle E_{Di} \rangle + \frac{h\nu_i}{\exp\left(\dfrac{h\nu_i}{kT_{\mathrm{n}}}\right) - 1} \tag{9.17}$$

where $\langle E_{Di} \rangle$ is the energy flow through the Fröhlich system or through the oscillating unit as is derived in the Appendix 13.11 (Eqs (13.93) and (13.94)) and T_{n} is the temperature connected with noise.

The entropy $\langle I(B) \rangle$ for a unit frequency bandwidth may be given by the relation [158]

$$\langle I(B) \rangle = \frac{1}{k \ln 2} \int_0^{T_{\mathrm{s}i}} \frac{\mathrm{d}\langle E_{\mathrm{n}i}(T) \rangle}{T} \, . \tag{9.18}$$

As the conditional information entropy $\langle I(A/B) \rangle$ might be given by [158]

$$\langle I(A/B) \rangle = \frac{1}{k \ln 2} \int_0^{T_{\mathrm{n}}} \frac{\mathrm{d}\langle E_{\mathrm{n}i}(T) \rangle}{T} \tag{9.19}$$

the capacity of the channel has the form [158]

$$C_i = \frac{1}{k \ln 2} \int_{T_{\mathrm{n}}}^{T_{\mathrm{s}i}} \frac{\mathrm{d}\langle E_{\mathrm{n}i}(T) \rangle}{T} \, . \tag{9.20}$$

Substituting relation given by Eq. (9.7) for $\langle E_{\mathrm{n}i}(T) \rangle$ into Eq. (9.20) we get (Appendix 13.10 [158])

$$C_i = \lg\left\{1 + \frac{\langle E_{Di} \rangle}{h\nu_i}\left[1 - \exp\left(-\frac{h\nu_i}{kT}\right)\right]\right\} - \frac{h\nu_i}{kT} \frac{1}{\ln 2 \left[\exp\left(\dfrac{h\nu_i}{kT}\right) - 1\right]}$$
$$+ \left[\frac{\langle E_{Di} \rangle}{h\nu_i} + \frac{1}{\exp\left(\dfrac{h\nu_i}{kT}\right) - 1}\right] \lg\left\{1 + \frac{h\nu_i\left[\exp\left(\dfrac{h\nu_i}{kT}\right) - 1\right]}{\langle E_{Di} \rangle\left[\exp\left(\dfrac{h\nu_i}{kT}\right) - 1\right] + h\nu_i}\right\} \tag{9.21}$$

where $T = T_n$. We will use Eq. (9.21) to analyze the communication channel between two Fröhlich coherent systems or oscillating units.

We will evaluate the capacity of the channel as a function of the distance between them (i.e. as a function of the channel length). A channel for the information transfer mediated by the radiation component of the electromagnetic field in the equatorial region of a dipole oscillator was analyzed in [161]. The energy flow through the Fröhlich system or the oscillating unit $\langle E_{Di} \rangle = \langle W_{iP} \rangle$, is derived in the Appendix 13.11. Figs 9.6a, b show the capacity of the "radiation field" channel for a great flow of energy ($p_i = e\ell$,

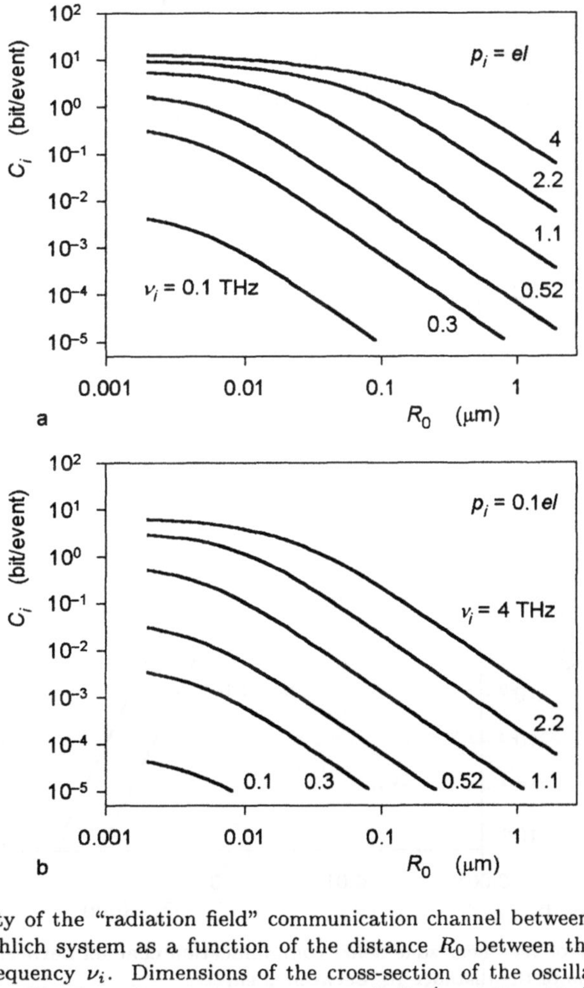

Fig. 9.6 Capacity of the "radiation field" communication channel between two oscillating units of the Fröhlich system as a function of the distance R_0 between them for different values of the frequency ν_i. Dimensions of the cross-section of the oscillating unit: $x = 5\,\text{nm}$, $z = 10\,\text{nm}$. The Poynting vector parameter: a) $G_i \nu_1^4 = 5.31 \times 10^{-24}$ VA ($p_i = e\ell$), b) $G_i \nu_1^4 = 5.31 \times 10^{-26}$ VA ($p_i = 0.1\,e\ell$).

where e is the elementary charge and $\ell = 10\,\mathrm{nm}$) and for a small energy flow ($p_i = 0.1\,e\ell$), respectively. The parameter of the curves is frequency. The relative permittivity of the medium around the oscillating units is assumed to be 1. The capacity of the channel depends on frequency. At the low freqencies the capacity of the channel is very small and the exponential decrease begins at a very small distance (of about 10 nm). The smaller the excitation (i.e. the smaller p_i), the greater the information loss at the low frequencies. At the high frequencies (2.2 THz and 4 THz) the capacity of the channel is great and the exponential decrease appears at a distance greater than about 0.1 μm.

Besides the radiation component the near zone components of the electric

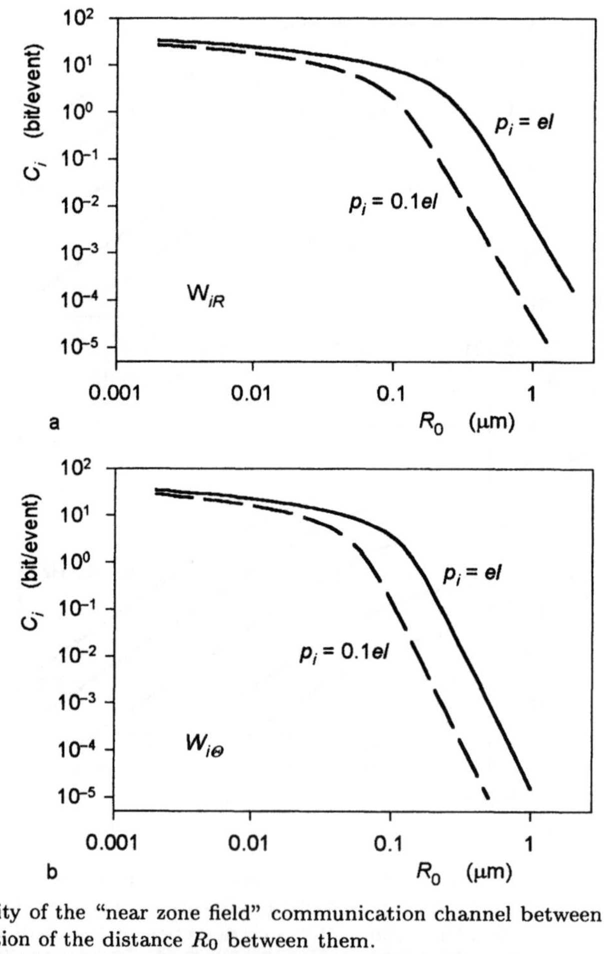

Fig. 9.7 Capacity of the "near zone field" communication channel between two oscillating units as a function of the distance R_0 between them.

a) Channel in the direction of the dipole axis ($\langle W_{iR} \rangle$).

b) Channel in the equatorial plane perpendicular to the dipole axis ($\langle W_{i\Theta} \rangle$).

field can mediate information transfer. Fig. 9.7a shows the capacity of the communication channel in the direction of the dipole axis (for $\langle E_{Di} \rangle = \langle W_{iR} \rangle$ derived in Appendix 13.11) [161]. Fig. 9.7b shows the capacity of the communication channel in the equatorial plane in the direction perpendicular to the dipole axis (for $\langle E_{Di} \rangle = \langle W_{i\Theta} \rangle$; Appendix 13.11) [161]. Within a distance smaller than about 0.1 μm the values of the capacity of the "near zone field" channels are several times greater than the capacity of the "radiation field" channel. The values of the capacity of the "near zone field" channels are nearly the same for different frequencies and start to decrease at smaller distances (and the decrease is much steeper) than the values of the "radiation field" channel. The high capacity of the "near zone field" channel in the equatorial region is limited to smaller distance than the capacity of the "near zone field" channel along the dipole axis. It may be concluded that the "near zone field" channels are dominant within a distance smaller than about 0.1 μm and the "radiation field" channel in the distance greater than about 0.1 μm.

Within a distance smaller than 0.1 μm the transfer of information is independent of frequency. At a distance greater than about 0.1 μm the high frequency spectral components of the endogenous electromagnetic field may have a fundamental role in transfer of information and in maintenance of coherence (e.g. in living cells) as follows from Figs 9.6 and 9.7. The distance between the protein molecules in a cell membrane may be of the order of 0.01–0.1 μm. At a distance of 0.1 μm the information transfer at 0.1 THz is suppressed. In contrast the information transfer at 4 THz is nearly unchanged (for the great energy flow – Fig. 9.6a). If the high frequency normal modes are not excited the information transfer between the coherent vibration systems or between the osillating units could not be sufficient to maintain coherence. Similar effects may result from the increased distance between the vibration systems (e.g. between protein molecules). The greater the distance the smaller the information transfer mediated by the endogenous electromagnetic field especially by the low frequency spectral lines. If the distance between two Fröhlich systems or oscillating units increases a sudden "cut off" of the information transfer may appear as a result of steep decrease of channel capacity as follows from Figs 9.6 and 9.7.

10. Interaction Between Vibration Systems

10.1 Frequencies of Interacting Oscillators

Elementary analysis of interacting identical oscillators with no damping is given in Appendix 13.12. In this part we will apply the concept of the interacting oscillators to the Fröhlich systems. First of all we will derive relations for frequencies of two interacting oscillators.

We adopt a model consisting of two large vibration systems (system 1 and system 2) at the distance R considerably larger than their linear dimensions. The systems are capable of dipole vibrations with the angular frequencies ω_1 and ω_2. The vibration systems interact with one another. We will assume that the interaction between these large systems arises from the interaction between their dipoles. The restoring forces $\mathcal{F}_1$, $\mathcal{F}_2$ of the oscillation system 1 and of the oscillation system 2 and the interaction forces $\mathcal{F}_{12}$, $\mathcal{F}_{21}$ between the systems are given by

$$\mathcal{F}_1 = -f_1 q_1 , \quad \mathcal{F}_2 = -f_2 q_2 , \quad \mathcal{F}_{12} = \gamma_1 q_2 , \quad \mathcal{F}_{21} = \gamma_2 q_1 , \tag{10.1}$$

respectively. In Eq. (10.1) q_1, q_2 are the displacements from the equilibrium positions; f_1, f_2 are the elastic force constants; and γ_1, γ_2 are the interaction coefficients describing the interaction between the systems. For harmonic oscillators with the masses m_1 and m_2 and with no damping the elastic force constants are $f_1 = m_1\omega_1^2$ and $f_2 = m_2\omega_2^2$. The differential equations of both oscillators are

$$m_1 q_1^{(2)} + m_1\omega_1^2 q_1 = \gamma_1 q_2 , \quad m_2 q_2^{(2)} + m_2\omega_2^2 q_2 = \gamma_2 q_1 . \tag{10.2}$$

From the first equation in (10.2) we determine the expression for q_2

$$q_2 = \frac{m_1}{\gamma_1} q_1^{(2)} + \frac{m_1\omega_1^2}{\gamma_1} q_1$$

and substitute it into the second equation in (10.2). We get

$$q_1^{(4)} + \left(\omega_1^2 + \omega_2^2\right) q_1^{(2)} + \left(\omega_1^2\omega_2^2 - \frac{\gamma_1\gamma_2}{m_1 m_2}\right) q_1 = 0 . \tag{10.3}$$

We assume the solution of Eq. (10.3) in the form $q_1 = A\exp(\mathrm{i}\omega t)$. One gets from Eq. (10.3)

$$\omega_{\pm}^2 = \frac{\omega_1^2+\omega_2^2}{2} \pm \sqrt{\left(\frac{\omega_1^2-\omega_2^2}{2}\right)^2 + \frac{\gamma_1\gamma_2}{m_1 m_2}}\,. \tag{10.4}$$

We have to emphasize that ω_1 and ω_2 in Eq. (10.4) represent the properties of the system 1 and of the system 2, respectively.

Eq. (10.4) is a well known formula for the frequencies of a joint system of two interacting oscillators. But only a simplified example is considered. Each oscillating system has only one vibration mode. If each oscillating system has several vibration modes a complicated spectrum of frequencies is generated as a result of interaction. The interaction terms $\gamma_1 q_2$ and $\gamma_2 q_1$ in Eqs (10.1) and (10.2) may also depend on q_1 and q_2, respectively, i.e. $\gamma_1 = \gamma_1(q_1)$ and $\gamma_2 = \gamma_2(q_2)$. The interaction coefficients can also depend on frequency and, therefore, the coefficients γ_{1+}, (γ_{2+}) at the frequency ω_+ are not equal to γ_{1-}, (γ_{2-}) at the frequency ω_-. In Eq. (10.4) we have to write $\gamma_{1\pm}$ and $\gamma_{2\pm}$ instead of γ_1 and γ_2.

From Eq. (10.4) we may determine condition for $\omega_- = \omega_+$. If $\omega_1 = \omega_2$ then the condition $\omega_+ = \omega_-$ is satisfied only for $\gamma_1\gamma_2 = 0$, i.e. if the systems do not interact. If $\omega_1 \neq \omega_2$ then the discriminant of Eq. (10.4) has to be zero. We get

$$\left(\frac{\omega_1^2-\omega_2^2}{2}\right)^2 = -\frac{\gamma_1\gamma_2}{m_1 m_2}\,. \tag{10.5}$$

As $m_1 > 0$ and $m_2 > 0$ then $\gamma_1\gamma_2 < 0$ and, therefore, in an isotropic linear medium the condition of equal frequencies $\omega_+ = \omega_-$ may be satisfied if

$$\gamma_1 = \pm\mathrm{i}|\gamma_1|\,, \quad \gamma_2 = \pm\mathrm{i}|\gamma_2| \quad \text{or} \quad \gamma_1 < 0\,, \quad \gamma_2 > 0 \quad \text{(or vice versa)} \tag{10.6}$$

and besides this condition the interaction coefficients γ_1, γ_2, the frequencies ω_1, ω_2, and the masses m_1, m_2, must have convenient values. This extraordinary condition can be fulfilled for two frequencies in special cases. In a dispersive and in a dissipative medium where $\gamma_{1+} \neq \gamma_{1-}$ and $\gamma_{2+} \neq \gamma_{2-}$ the equality condition for the frequencies ω_- and ω_+ can be satisfied only in an artificial case when $\gamma_{1+}\gamma_{2+} = \gamma_{1-}\gamma_{2-}$ as follows from Eqs (10.4) and (10.5). If γ_{1+}, γ_{1-} and γ_{2+}, γ_{2-} depend on q_1 and q_2, respectively, then equality of frequencies $(\omega_+ = \omega_-)$ may be satisfied under extraordinary conditions. The vibration systems may have more frequencies than two (i.e. more frequencies than ω_1 and ω_2). Convenient values of masses of the oscillators and extraordinary dispersive, dissipative, and non-linear properties of the medium between the oscillation systems (corresponding to the excitation of the oscilators) are very likely required to satisfy the equality condition of the frequencies of the interacting systems $\omega_i = \omega_j = \omega_k = \dots$. Nevertheless, such conditions are impossible in ordinary systems.

The differential equations given by (10.2) do not contain damping. The damping term representing energy leakage to the heat bath may introduce a new "coordinate" but as long as the system is linear no principle alterations in behaviour of the system may be expected.

10.2 Interaction Forces and Energy

The vibration systems are polar and interaction between them depends mainly on the electric field generated by the electric dipoles or multipoles. The intensity of the electric field generated by an electric dipole is given by the well known relations given in Appendix 13.7 – Eqs. (13.63). The relations for the intensity of the electric field contain terms which depend on the distance as $1/R^3$, $1/R^2$, and $1/R$. At a short distance the value of the term $1/R^3$ is greater than the values of the terms $1/R^2$ and $1/R$. Using the free space value of the permittivity ($\varepsilon = \varepsilon_0 = 8.854\times10^{-12}\,\mathrm{Fm}^{-1}$) and of the permeability ($\mu = \mu_0 = 1.257\times\,10^{-6}\,\mathrm{Hm}^{-1}$) we find for a frequency smaller than about 50 THz and for the distance R smaller than 1 μm that the absolute values of the second term in brackets on the right hand side of the first equation in (13.63) and of the second and of the third terms of the second equation in (13.63) are smaller than the first terms. The small terms may be neglected and only the first bracket terms on the right hand sides of the relations given in (13.63) remain.

The potential energy of a dipole in the external electric field is given by the scalar product $-\boldsymbol{p}_1 \cdot \boldsymbol{E}$ where $\boldsymbol{p}_1$ is the dipole moment and $\boldsymbol{E}$ is the intensity of the electric field (in this case generated by another dipole). Using Eqs (13.63) the "effective" intensity of the electric field acting on the dipole 1 is

$$E \simeq \frac{p_2}{\varepsilon R^3} f(\Theta) \tag{10.7}$$

where p_2 is the dipole moment of the vibration system 2 and $f(\Theta)$ is a symbol for the equivalent function representing various space orientations of the dipoles.

We assume that the interaction coefficients γ_1 and γ_2 in Eq. (10.4) may be approximated by

$$\gamma_1 = \alpha\frac{p_2}{\varepsilon R^3} f(\Theta)\,, \quad \gamma_2 = \alpha\frac{p_1}{\varepsilon R^3} f(\Theta) \tag{10.8}$$

where α is the coefficient of proportionality.

The interaction energy I' is defined as a difference between the energy $U(R)$ of both systems at a distance R and the energy $U(\infty)$ (if exists) at

$R \to \infty$ (Appendix 13.12), i.e.

$$I' = U(R) - U(\infty) . \tag{10.9}$$

The interaction force is given by the relation

$$\mathcal{F} = -\frac{\mathrm{d}I'}{\mathrm{d}R} . \tag{10.10}$$

We will evaluate the interaction energy related to one system. Therefore, we will take one half of the interaction energy defined by Eq. (10.9), i.e.

$$I = \frac{I'}{2} = \frac{1}{2}\left[U(R) - U(\infty)\right] . \tag{10.11}$$

If the interaction force in Eq. (10.10) is greater than zero ($\mathcal{F} > 0$) the interaction results in repulsion; if $\mathcal{F} < 0$ then the interaction results in attraction between the vibration systems.

Interaction forces betwen coherent oscillators were analyzed by Fröhlich in [75,79,160]. Analyses of the interaction forces between models of the coherent vibration systems are given in [129,157,161].

10.3 Quantum Case of Interaction

We will analyze interaction of two systems where each of them has only one harmonic oscillator. An extreme quantum case occurs if $kT \ll h\nu$. The interaction energy of one system given generally by Eq. (10.11) may be written in the form

$$I = \frac{h}{2}\left[n_-\nu_- + n_+\nu_+ - n_1\nu_1 - n_2\nu_2\right] \tag{10.12}$$

where n_1 (n_2) is the occupation number of the normal mode with the frequency ν_1 (ν_2) when the coherent vibration systems do not interact; n_- (n_+) is the occupation number of the normal mode with the frequency ν_- (ν_+) of the joint interacting vibration systems. After substitution from Eq. (10.4) for ν_- and for ν_+ into Eq. (10.12) one gets

$$\begin{aligned} I = {} & \frac{h}{2}n_-\sqrt{\frac{\nu_1^2+\nu_2^2}{2}}\left\{1-\left[\left(\frac{\nu_1^2-\nu_2^2}{\nu_1^2+\nu_2^2}\right)^2+\frac{\gamma_{1-}\gamma_{2-}}{(2\pi)^4 m_1 m_2}\left(\frac{2}{\nu_1^2+\nu_2^2}\right)^2\right]^{\frac{1}{2}}\right\}^{\frac{1}{2}} \\ & +\frac{h}{2}n_+\sqrt{\frac{\nu_1^2+\nu_2^2}{2}}\left\{1+\left[\left(\frac{\nu_1^2-\nu_2^2}{\nu_1^2+\nu_2^2}\right)^2+\frac{\gamma_{1+}\gamma_{2+}}{(2\pi)^4 m_1 m_2}\left(\frac{2}{\nu_1^2+\nu_2^2}\right)^2\right]^{\frac{1}{2}}\right\}^{\frac{1}{2}} \\ & -\frac{h}{2}\left(n_1\nu_1+n_2\nu_2\right) . \end{aligned} \tag{10.13}$$

Eq. (10.13) will be analyzed for a non-resonant case (i.e. $\nu_1 \neq \nu_2$) and for a resonant case ($\nu_1 = \nu_2$).

10.3.1 Non-Resonant Case

We will find a solution for $\nu_1 \neq \nu_2$ (i.e. in the non-resonant case). Replacing γ_{1-}, γ_{2-} and γ_{1+}, γ_{2+} in Eq. (10.13) by the expressions given by Eq. (10.8), and using the Binomial Theorem (twice) one gets from Eq. (10.13) (after neglecting the higher order terms)

$$I = \frac{h}{2}\sqrt{\frac{\nu_1^2+\nu_2^2}{2}}\left[n_- + n_+ - B(n_- - n_+) - \frac{C}{R^6}\left(\frac{n_-}{\varepsilon_-^2} - \frac{n_+}{\varepsilon_+^2}\right)\right]$$
$$-\frac{h}{2}\left(n_1\nu_1 + n_2\nu_2\right) \tag{10.14}$$

where

$$B = \frac{1}{2}\frac{\nu_1^2-\nu_2^2}{\nu_1^2+\nu_2^2}\,, \quad C = \frac{1}{\nu_1^4-\nu_2^4}\left[\frac{\alpha\sqrt{p_1p_2}f(\Theta)}{(2\pi)^2\sqrt{m_1m_2}}\right]^2 \tag{10.15}$$

and the symbols ε_- and ε_+ denote the permittivity at the frequency ν_- and ν_+, respectively. The coefficient C given by Eq. (10.15) is inversely proportional to the difference $\nu_1^4 - \nu_2^4$ and steeply increases for $\nu_2 \to \nu_1$.

Eq. (10.14) holds for

$$1 \gg \sqrt{\left(\frac{\nu_1^2-\nu_2^2}{\nu_1^2+\nu_2^2}\right)^2 + \frac{\gamma_{1\pm}\gamma_{2\pm}}{(2\pi)^4 m_1m_2}\left(\frac{2}{\nu_1^2+\nu_2^2}\right)^2}$$
$$\left(\frac{\nu_1^2-\nu_2^2}{\nu_1^2+\nu_2^2}\right)^2 \gg \frac{\gamma_{1\pm}\gamma_{2\pm}}{(2\pi)^4 m_1m_2}\left(\frac{2}{\nu_1^2+\nu_2^2}\right)^2 . \tag{10.16}$$

The interaction energy given by Eq. (10.14) depends on the frequencies of the oscillators, on their excitations (i.e. on the occupation numbers n_-, n_+, n_1, n_2) and on the permittivity of the surrounding medium. In particular we point out that the interaction energy given by Eq. (10.14) depends on the permittivity of the medium between the vibration systems. The dispersion term depending on the permittivity of the ambient medium steeply decreases with distance as R^{-6}.

10.3.2 Resonant Case

For a resonant case $\nu_1 = \nu_2$ and $\nu_1^2 - \nu_2^2$ in Eq. (10.13) is zero. Substituting formulae from (10.8) for $\gamma_{1\pm}$ and $\gamma_{2\pm}$ one gets from Eq. (10.13) the relation

$$I = \frac{h}{2}\left[\left(n_- + n_+ - n_1 - n_2\right)\nu_1 - \frac{D}{R^3}\left(\frac{n_-}{\varepsilon_-} - \frac{n_+}{\varepsilon_+}\right)\right] \tag{10.17}$$

where

$$D = \frac{1}{2\nu} \frac{\alpha\sqrt{p_1 p_2} f(\Theta)}{(2\pi)^2 \sqrt{m_1 m_2}} \tag{10.18}$$

The first term on the right hand side of Eq. (10.17) depends on the excitation level of the oscillators (i.e. on the occupation numbers). If $n_- = n_+ = n_1 = n_2$ the first term vanishes. The second term in Eq. (10.17) depends on the dispersion of the ambient medium. This dispersion term is a function of the distance and varies as R^{-3}. In a non-resonant state the dispersion term depends on distance as R^{-6} as follows from Eq. (10.14). At resonance the decrease of the interaction forces with distance is much smaller than in a non-resonant state as follows from comparison of the terms C/R^6 and D/R^3 in Eqs (10.14) and (10.17), respectively. Equations (10.14) and (10.17) reveal a possibility of the frequency selective long range forces between the vibration systems provided the permittivity of the medium between them shows appropriate dispersion in the frequency region in question. The frequencies ν_- and ν_+ depend on the interaction between the vibration systems, i.e. on the intensity of the electric field generated by the vibration systems.

We will analyze the interaction energy given by Eq. (10.17) under the condition $n_- = n_+ = n_1 = n_2$. From Eq. (10.17) one gets the dispersion term of the interaction energy in the form

$$I = -\frac{h}{2} D \frac{n_1}{R^3} \left(\frac{1}{\varepsilon_-} - \frac{1}{\varepsilon_+} \right) . \tag{10.19}$$

The interaction force between the systems is given by

$$\mathcal{F} = -2\frac{\mathrm{d}I}{\mathrm{d}R} = -hD\frac{3n_1}{R^4} \left(\frac{1}{\varepsilon_-} - \frac{1}{\varepsilon_+} \right) . \tag{10.20}$$

If the permittivity of the medium increases with the increasing frequency then

$$\varepsilon_- < \varepsilon_+ , \quad \frac{1}{\varepsilon_-} > \frac{1}{\varepsilon_+} , \quad \mathcal{F} = -2\frac{\mathrm{d}I}{\mathrm{d}R} < 0$$

and the dispersion term causes attraction.

If the permittivity decreases with the increasing frequency, then

$$\varepsilon_- > \varepsilon_+ , \quad \frac{1}{\varepsilon_-} < \frac{1}{\varepsilon_+} , \quad \mathcal{F} = -2\frac{\mathrm{d}I}{\mathrm{d}R} > 0$$

and the dispersion term represents a repulsive interaction force.

The interaction force derived from Eq. (10.19) may be attractive or repulsive which depends on the sign of $\mathrm{d}\varepsilon/\mathrm{d}\nu$ (for a monotonous dependence). In the region of anomalous dispersion where permittivity decreases with increasing frequency the interaction forces may be repulsive. Anomalous dispersion occurs in the region of resonant absorption of electromagnetic waves. The wave numbers of the bending and stretching vibrations of water molecules are about $1600\,\mathrm{cm}^{-1}$ (i.e. $4.8\times10^{13}\,\mathrm{Hz}$) and $4000\,\mathrm{cm}^{-1}$ (i.e. $1.2\times10^{14}\,\mathrm{Hz}$), respectively.

The resonant interaction mediated by the generated electric field is a long range interaction which may have fundamental biological significance.

10.4 Classical Case of Interaction

An extreme classical case of interaction occurs if $kT \gg h\nu$. In the isothermal isometric system the free energy is given by the relation $F = U - TS$ where U is the total energy and S is the entropy. Entropy of a system in certain thermodynamic state is given by the Boltzmann relation $S = k \ln P$ where P is the probability of the occurence of the thermodynamic state. The classical statistics of Boltzmann yields for the entropy of an oscillator in the one-dimensional case (when only one position coordinate exists) the relation [35]

$$S = \frac{U}{T} + kN \ln \frac{\sigma}{h} \,. \tag{10.21}$$

and the free energy is given by

$$F = U - ST = -kTN \ln \frac{\sigma}{h} \tag{10.22}$$

where σ is the partition sum. The partition sum σ of an oscillator in thermodynamic equilibrium with one position coordinate is given by the relation [35]

$$\sigma = h \sum_{n=0}^{+\infty} \exp\left(-\frac{nh\nu}{kT}\right) = h\left[1 - \exp\left(-\frac{h\nu}{kT}\right)\right]^{-1} \approx h\frac{kT}{h\nu} \,. \tag{10.23}$$

The interaction energy ($I = I'/2$) of two oscillators in thermodynamic equilibrium is given by

$$\begin{aligned} I &= \frac{1}{2}kTN\left[-\ln\frac{kT}{h\nu_-} - \ln\frac{kT}{h\nu_+} + \ln\frac{kT}{h\nu_1} + \ln\frac{kT}{h\nu_2}\right] \\ &= \frac{1}{2}kTN \ln \frac{\nu_-\nu_+}{\nu_1\nu_2} \,. \end{aligned} \tag{10.24}$$

If the energy of an oscillator is higher than its energy in thermodynamic equilibrium ($n = n_0 + \Delta n$) we make use of the "shifted" distribution and the "shifted" partition sum. The probability of energy level given by the occupation number $n + \Delta n$ is $\exp[-(n + \Delta n)\,h\nu/kT]$. The partition sum σ is given by the relation

$$\sigma = h \sum_{n_0=0}^{\infty} \exp\left[-(n_0 + \Delta n)\,\frac{h\nu}{kT}\right]$$
$$= h \exp\left(-\frac{\Delta n h\nu}{kT}\right)\left[1 - \exp\left(-\frac{h\nu}{kT}\right)\right]^{-1} = h \exp\left(-\frac{\Delta n h\nu}{kT}\right)\frac{kT}{h\nu}\,. \tag{10.25}$$

The interaction energy of two oscillators is given by

$$I = \frac{h}{2} N \left[\Delta n_- \nu_- + \Delta n_+ \nu_+ - \Delta n_1 \nu_1 - \Delta n_2 \nu_2 + \frac{kT}{h} \ln \frac{\nu_- \nu_+}{\nu_1 \nu_2}\right] \tag{10.26}$$

The last term on the right hand side of Eq. (10.26) is the thermodynamic equilibrium term. The first four terms are connected with the state far from thermodynamic equilibrium.

10.5 Approximate Relations for Interaction Energy

In the previous section interaction between two oscillators was analyzed. A vibration system may have a great number of normal modes with different frequencies. Each normal mode of one system interacts with all the normal modes of the other system. Therefore, the spectral line of each normal mode of the joint vibration system is split up into a great number of particular frequency lines. The number of particular frequency lines is in relation to the number of the normal modes in the vibration systems before interaction.

We assume that the interacting systems are identical and that only the resonant interaction is significant. Therefore, we assume that each normal mode is split up into two frequency lines. For the interaction energy in the extreme quantum case ($h\nu \gg kT$) one may write

$$I = \frac{h}{2}\left[\left(\sum_k n_k \nu_k\right)_{\mathrm{I}} - \left(\sum_i n_i \nu_i\right)_1 - \left(\sum_j n_j \nu_j\right)_2\right] \tag{10.27}$$

where the indices I, 1, and 2 denote the joint vibration system (i.e. the two interacting systems), the vibration system 1, and the vibration system 2 (before interaction), respectively.

In the extreme classical case ($h\nu \ll kT$) the interaction energy is given by

$$I = \frac{h}{2}N\left[\left(\sum_k \Delta n_k \nu_k\right)_{\mathrm{I}} - \left(\sum_i \Delta n_i \nu_i\right)_1 - \left(\sum_j \Delta n_j \nu_j\right)_2 \right.$$
$$\left. + \left(\frac{kT}{h}\sum_k \ln\frac{h\nu_k}{kT}\right)_{\mathrm{I}} - \left(\frac{kT}{h}\sum_i \ln\frac{h\nu_i}{kT}\right)_1 - \left(\frac{kT}{h}\sum_j \ln\frac{h\nu_j}{kT}\right)_2\right] . \qquad (10.28)$$

Equations (10.27) and (10.28) may be used for numerical evaluation of the interaction energy. These relations enable us to analyze interactions between the Fröhlich coherent systems under different conditions. In particular, we can examine the effect of excitation and coupling to the heat bath.

The Fröhlich interaction forces are connected with the coherent systems far from thermodynamic equilibrium. The energy condensed in some vibration modes is much greater than the energy of these modes in thermodynamic equilibrium. The interaction forces between the Fröhlich coherent vibration systems depend on the difference of the condensed energy in the interacting systems as follows from Eqs 10.14 and 10.17 and may be, therefore, much greater than van der Waals interaction forces which occur in thermodynamic equilibrium. In particular, the interaction forces may be very strong in the resonant case as may be assessed from Eq. 10.17.

Electrostatic or quasi-electrostatic forces are short range forces. Static electric charges are screened by free charges in the surrounding medium (Appendix 13.13). The concentration of free ions in the cytosol and in the liquid medium surrounding cells may be about 10^{-1} M [1]. Free ions with this concentration are capable of the screening of static charges within a distance considerably smaller than 10 nm as is derived in Appendix 13.13. The static potential approaches zero over a distance which is referred to as the Debye thickness. Electrostatic forces are significant within a distance at least one order smaller than the forces based on the oscillating electric field with a sufficiently high frequency. The Fröhlich forces are mediated by oscillating electric fields.

11. Heat Bath Coupling Effects

11.1 Decrease of Energy Condensation

The interaction of the Fröhlich vibration system with the heat bath has a crucial effect on the energy transfer and energy condensation in the vibration systems. The coupling of the vibration system to the heat bath determines the transition probabilities of the energy transfer. The χ term in the Fröhlich rate equation (5.11) represents the energy transfer between normal modes with different frequencies. The participation of the heat bath makes possible the energy transfer. The χ term represents a non-linear process: interaction between two modes of the system and the heat bath which balances the difference of energy quanta. Without this non-linear coupling to the heat bath the χ energy transfer is not possible and the energy condensation is missing as follows from Eqs (4.24) and (5.11) which describe the Fröhlich model. On the other hand for a strong coupling to the heat bath energy losses from the vibration system are high and the condensed energy is low. The amount of the energy condensed in the vibration system may be important for living processes. Conditions for energy condensation are, therefore, fundamental.

We will analyze the effect of the heat bath on the energy condensation and on the interaction between vibration systems. We use the numerical method described in the Chapter 5 [114]. The energy saturation condition is included in the computation algorithm. The χ transition probability is independent of the occupation numbers and the Φ transition probability depends on the occupation number (i.e. $\Phi = \Phi(n_i)$) as is given by Eq. (5.39). Most of the energy above a limit corresponding to certain occupation number $n_{\max}$ (i.e. above a point of "saturation") is transferred to the heat bath. In our case we use $n_{\max} = 300$ and $n_{\max} = 3500$. The results given in this chapter represent the steady state conditions.

We will examine simple models of vibration systems in the frequency range from 10^{10} Hz to 10^{13} Hz (similar models are used in section 5.5). With respect to the χ_{ij} values the model systems consist of two group of modes: the low frequency group of modes and the high frequency group of modes. The χ transition probabilities in both these groups are the same ($\chi_{ij} = \chi_a$ for $i < 4$, $j < 4$, and for $i \geq 4$, $j \geq 4$, where $i = 4$ denotes the quasi-lowest fre-

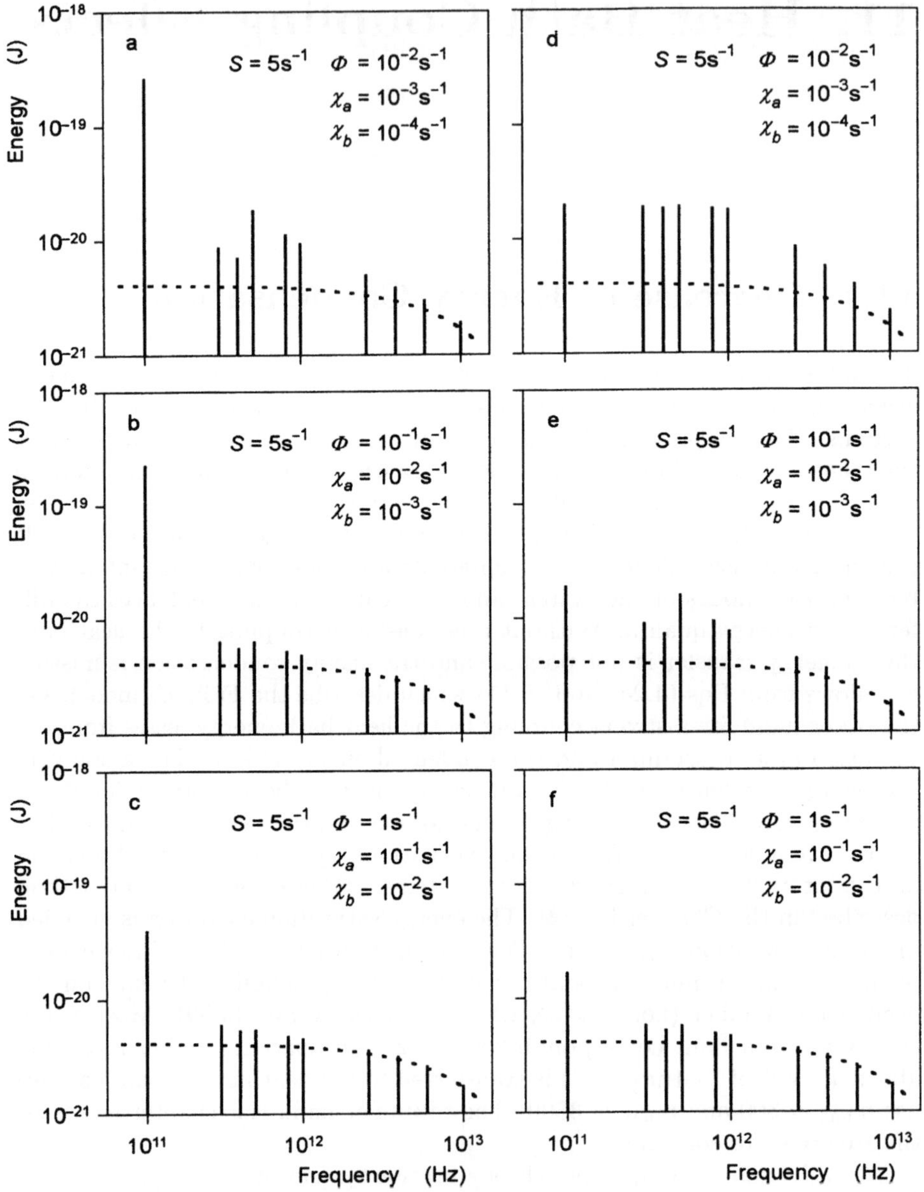

Fig. 11.1 Energy of the normal modes as a function of frequency. The dotted lines indicate the thermodynamic equilibrium values. The energy supply to all vibration modes is the same ($S = 5\,\mathrm{s}^{-1}$). Coupling to the heat bath is:

a) weak, b) medium, c) strong ($n_{\max} = 3500$),

d) weak, e) medium, f) strong ($n_{\max} = 300$).

quency mode) but the transition probabilities between the two groups of modes are considerably smaller ($\chi_{ij} = \chi_b = 0.1\chi_a$ for $i < 4,\ j \geq 4$ and $i \geq 4,\ j < 4$) than inside the groups. Figure 11.1 shows the energy in particular modes for different Φ_i and χ_{ij} transmission probabilities and for $n_{\max} = 3500$ and $n_{\max} = 300$. Similar analysis is given in [130].

Figs 11.1a and 11.1d show the vibration systems with the weak coupling to the heat bath ($\Phi_i = 10^{-2}\,\mathrm{s}^{-1}$, $\chi_a = 10^{-3}\,\mathrm{s}^{-1}$). The vibration systems are highly excited. The energy condensed in the lowest frequency mode is much greater than the energy in the thermodynamic equilibrium state which is a condition for giant coherent oscillations. The energy is condensed not only in the lowest frequency mode of the low frequency group of modes but also in the quasi-lowest frequency mode ν_4 (in the lowest frequency mode of the high frequency group of modes); the modes in its vicinity are considerably excited too. The model system with $n_{\max} = 3500$ (Fig. 11.1a) has great differences of the energy levels in the vibration modes with condensed energy. The model system with $n_{\max} = 300$ (Fig. 11.1d) has six excited vibration modes with nearly the same occupation numbers with frequencies from ν_1 to ν_6.

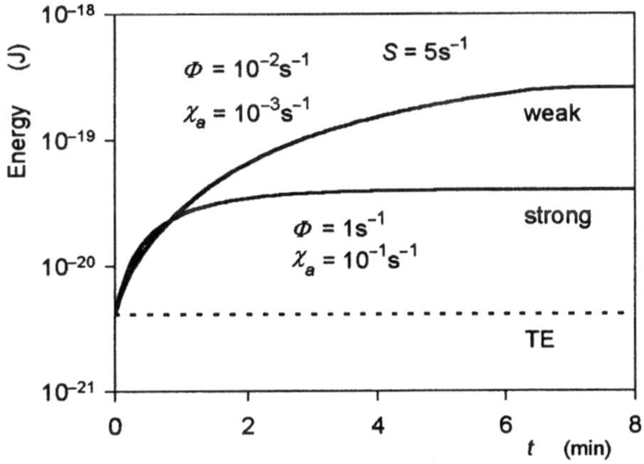

Fig. 11.2 Energy in the lowest frequency mode as a function of time. The same model of the vibration system as is described in Fig. 11.1 was used. The dotted line indicates the thermodynamic equilibrium values.

Figs 11.1b and 11.1e show the vibration systems with the medium coupling to the heat bath ($\Phi_i = 0.1\,\mathrm{s}^{-1}$, $\chi_a = 10^{-2}\,\mathrm{s}^{-1}$). The values of Φ and of χ are one order greater that those in Figs 11.1a and 11.1d. Fig. 11.1b shows a model where the energy in the quasi-lowest frequency mode is nearly in thermodynamic equilibrium ($n_{\max} = 3500$). Fig. 11.1e shows a model in which the great value of the condensed energy remains in the quasi-lowest frequency mode but the condensed energy is small in the vibration modes in its vicinity ($n_{max} = 300$). We note that the system with $n_{max} = 300$ needs smaller energy to reach the saturated state than the system with the high value of n_{max}. Figs 11.1c and 11.1f show vibration systems with the strong coupling to the heat bath ($\Phi_i = 1\,s^{-1}$, $\chi_a = 0.1\,s^{-1}$). Even the energy condensed in the lowest frequency mode is decreased. This decrease is less noticeable in the model for $n_{max} = 300$ (Fig. 11.1f).

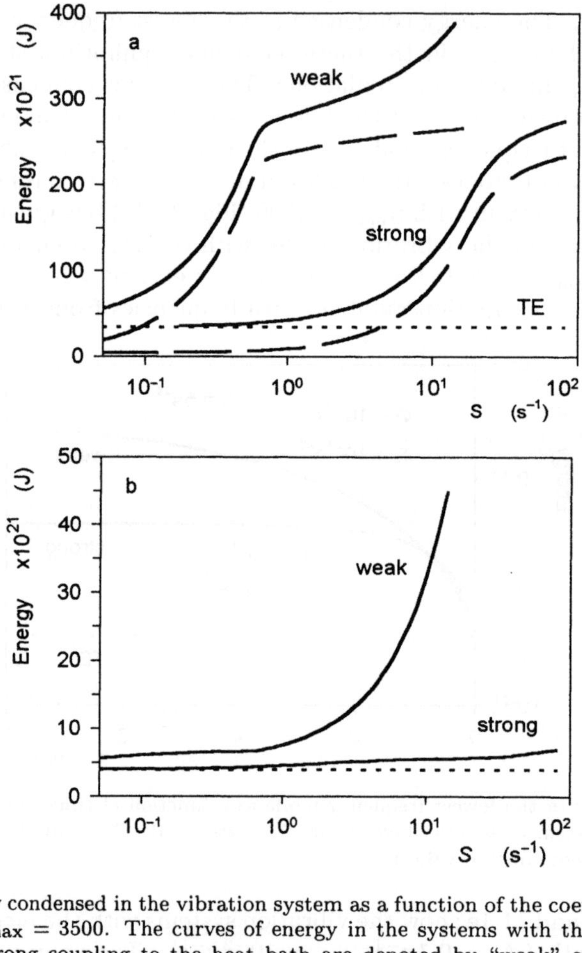

Fig. 11.3 Energy condensed in the vibration system as a function of the coefficient of energy supply S for $n_{max} = 3500$. The curves of energy in the systems with the weak coupling and with the strong coupling to the heat bath are denoted by "weak" and by "strong", respectively. The dotted lines indicate the thermodynamic equilibrium values: of the whole vibration system (a) and of the quasi-lowest frequency mode (b). The frequencies of normal modes are the same as the frequencies of the model in Fig. 11.1.

a) Total energy of the vibration system (the full curves) and energy condensed in the lowest frequency mode (the dashed curves).

b) Energy condensed in the quasi-lowest frequency mode (ν_4).

Fig. 11.2 shows the time dependence of the energy in the lowest frequency mode for the same vibration system as in Figs 11.1a and 11.1c (similar model was described in [130]). The energy versus time is a smooth monotonously increasing curve. In both cases shown in Fig. 11.2 the steady state conditions are reached within about 6 min from the beginning when the energy supply was switched on. The energy condensed in the lowest frequency mode in the system with the weak coupling to the heat bath reaches the maximum value

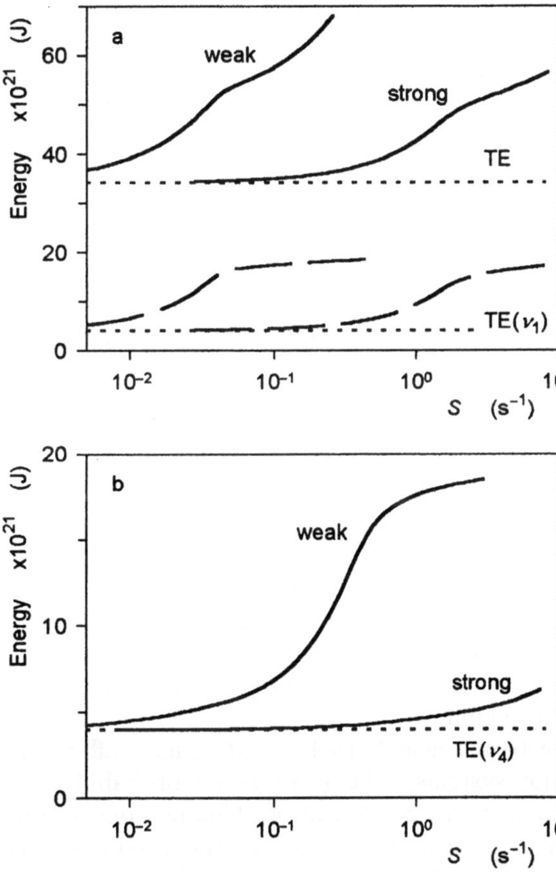

Fig. 11.4 Energy condensed in the vibration system as a function of S for $n_{\max} = 300$. The dotted lines indicate the thermodynamic equilibrium values: TE(ν_1) and TE denote the energy in the lowest frequency mode and in the whole vibration system, respectively (a) and TE(ν_4) the energy in the quasi-lowest frequency mode (b). The frequencies of normal modes are the same as the frequencies of the model in Fig. 11.1.

a) Total energy of the vibration system (the full curves) and the energy condensed in the lowest frequency mode (the dashed curves).

b) Energy condensed in the quasi-lowest frequency mode.

(i.e. the value corresponding to $n_{\max} = 3500$ of the occupation number) at about 5 min from the beginning. The majority of the energy over this maximal value is transferred to the heat bath. The energy condensed in the system with the strong coupling to the heat bath reaches the steady state at a shorter time but the energy is smaller than the energy in the system with the weak coupling to the heat bath (in this case $n_1 < n_{\max}$). The smaller the non-linear coupling to the heat bath, the greater the condensed energy, but the longer the time for reaching the steady state.

Figs 11.3 and 11.4 show the energy in the lowest frequency mode, the energy of the whole vibration system, and the energy in the quasi-lowest frequency mode as a function of the energy supply (of the coefficient of energy supply S). Similar dependence was analyzed in [129,157]. The $n_{\max}$ values are 3500 (Figs 11.3a, b) and 300 (Figs 11.4a, b). The vibration systems with the weak coupling to the heat bath ($\Phi_i = 10^{-2}\,\mathrm{s}^{-1}$, $\chi_a = 10^{-3}\,\mathrm{s}^{-1}$) have strong condensation of the energy at the low energy supply ($S \approx 1\,\mathrm{s}^{-1}$ for $n_{\max} = 3500$, and $S \approx 0.1\,\mathrm{s}^{-1}$ for $n_{\max} = 300$). The vibration systems with the strong coupling to the heat bath ($\Phi_i = 1\,\mathrm{s}^{-1}$, $\chi_a = 0.1\,\mathrm{s}^{-1}$) need the high energy supply for strong condensation ($S \approx 30\,\mathrm{s}^{-1}$ for $n_{\max} = 3500$ – Fig. 11.3a, and $S \approx 3\,\mathrm{s}^{-1}$ for $n_{\max} = 300$ – Fig. 11.4a). The vibration systems with the weak coupling to the heat bath are highly excited if S is greater than about $1\,\mathrm{s}^{-1}$ (Fig. 11.3a – $n_{\max} = 3500$) and if $S > 0.1\,\mathrm{s}^{-1}$ (Fig. 11.4a – $n_{\max} = 300$). For the same excitation the energy condensed in the vibration systems with the strong coupling to the heat bath does not differ considerably from the energy in thermodynamic equilibrium.

The energy condensed in the quasi-lowest frequency mode ν_4 as a function of the coefficient of the energy supply S is shown in Figs 11.3b and 11.4b. In a vibration system with the weak coupling to the heat bath energy condensed in the quasilowest frequency mode steeply increases at about $S = 10\,\mathrm{s}^{-1}$ – Fig. 11.3b ($n_{\max} = 3500$) and at about $S = 1\,\mathrm{s}^{-1}$ – Fig. 11.4b ($n_{\max} = 300$). In a vibration system with the strong coupling to the heat bath energy condensed in this mode is negligible.

The coupling to the heat bath has a dominant effect on energy condensation in vibration systems. The χ transition probability is one of the main factors determining energy condensation. Changes of the coupling to the heat bath result in great changes of the energy condensed in the vibration systems provided the energy supply remains unchanged.

11.2 Interaction Energy

The electromagnetic field generated by the polar vibration systems mediates an interaction between them. We make use of a simple model of two interacting vibration systems [129,157]. (Elementary analysis of the interaction is given in Appendix 13.12.) The assumed systems may be formed by vibrators in a cell. Two interacting systems form a joint system with the changed frequencies of the vibration modes. We assume that before interaction the frequencies of the vibration modes of both the systems are the same and that only the resonant interactions between the modes of equal frequencies are important. Interactions between the modes with different frequencies are neglected. The frequencies of the vibration modes of the joint system are given by Eq. (10.4). We will use an approximate relation

$$\nu_\pm^2 = \left(\frac{\nu_i^2 + \nu_j^2}{2}\right) \pm \sqrt{\left(\frac{\nu_i^2 - \nu_j^2}{2}\right)^2 + (K^2\nu_i\nu_j)^2} \tag{11.1}$$

where K is the modified interaction coefficient. The modified interaction term $K^2\nu_i\nu_j$ in Eq. (11.1) depends on the frequencies ν_i and ν_j of the modes explicitly (we assume that $\nu_i = \nu_j$). The interaction coefficients γ_1 and γ_2 given by Eq. (10.8) depend on the frequencies of oscillations through permittivity of the ambient medium. In the millimeter and in the submillimeter wave regions permittivity of water decreases with decreasing wave length and the interaction coefficients γ_1 and γ_2 increase which justifies the use of the modified interaction term. We assume that K is a constant for all vibration modes. Fig. 11.5 shows the energy in the normal modes of two joint interacting systems for $n_{\max} = 3500$ (Figs 11.5a, b) and for $n_{\max} = 300$ (Figs 11.5c, d). Two joint interacting systems in Fig. 11.5a, c and 11.5b, d have the weak and the strong coupling to the heat bath, respectively. The modified interaction coefficient is a function of the generated electromagnetic field which depends on the coupling to the heat bath, on the energy supply, and on the distance from the source. As the intensity of the generated electric field decreases with the increasing distance K decreases with the distance too. The weaker is the generated electromagnetic field the smaller is K. Figs 11.5a–d are evaluated for the same value of K. Therefore, the distances R between the systems shown in Figs 11.5a, b, c, and d are not the same.

For the interaction energy of two interacting vibration system an approximate relation is used in the form

$$I = \frac{h}{2}\left[\frac{1}{2}\left(\sum_k n_k\nu_k\right)_{\mathrm{I}} + \frac{1}{2}\left(\sum_\ell n_\ell\nu_\ell\right)_{\mathrm{I}} - \left(\sum_i n_i\nu_i\right)_0 - \left(\sum_j n_j\nu_j\right)_0\right] \tag{11.2}$$

where the indices I and 0 denote the state in interaction and before interaction, respectively.

The interaction energy of two interacting systems is given in Fig. 11.6 [129, 157] as a function of the modified interaction coefficient K. The interaction energy depends on the coupling to the heat bath. The systems with the weak coupling to the heat bath have great absolute values of the interaction energy (Fig 11.6 – the full curve denoted by "weak-weak"). The interaction energy is negative and its absolute value increases with increasing modified interaction coefficient K. The interaction force given by

$$-2\frac{\mathrm{d}I}{\mathrm{d}R} = -2\left(\frac{\mathrm{d}I}{\mathrm{d}K}\right)\left(\frac{\mathrm{d}K}{\mathrm{d}R}\right) \tag{11.3}$$

is attractive as follows from Fig. 11.6 (as K decreases with increasing distance R between the vibration systems).

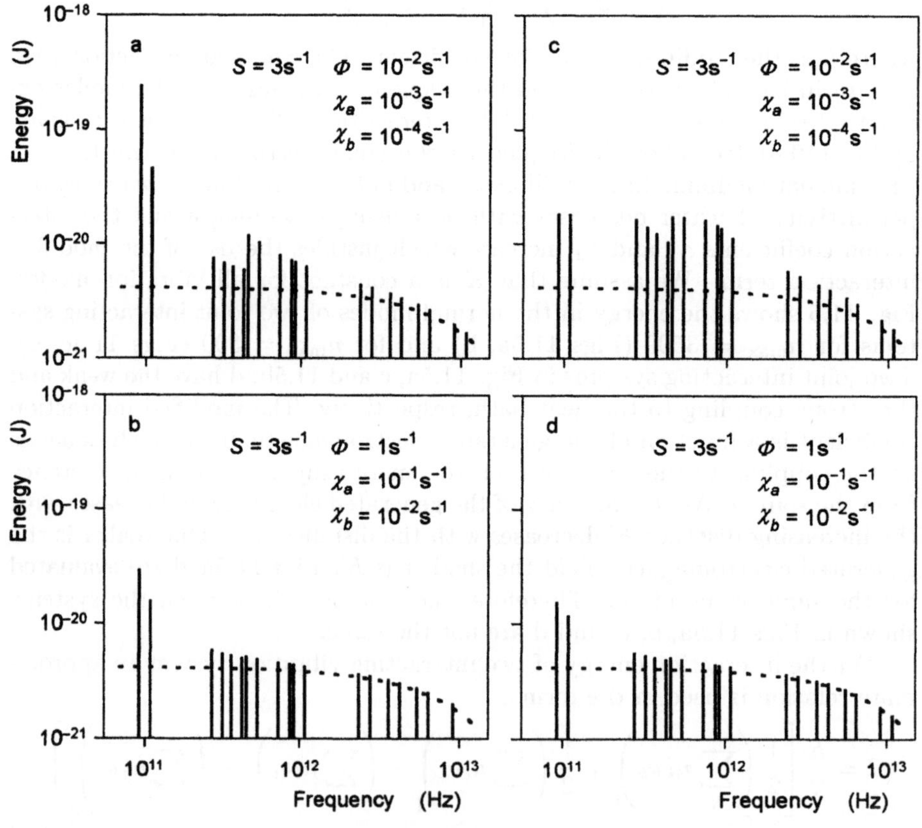

Fig. 11.5 Energy condensed in the normal modes of two interacting vibration systems as a function of frequency. $S = 3\,\mathrm{s}^{-1}$. The values of Φ_i and of χ_{ij} are the same as in Fig. 11.1a (the weak coupling to the heat bath – a, c) and in Fig. 11.1c (the strong coupling to the heat bath – b, d). The maximum value of the condensed energy is given by $n_{\max} = 3500$ (a, b) and by $n_{\max} = 300$ (c, d).

The vibration systems with the strong coupling to the heat bath have very small absolute values of the interaction energy (Fig. 11.6 – the dashed curve denoted by "strong-strong") which are about two orders smaller than those of the systems with the weak coupling to the heat bath ("weak-weak"). The attractive forces are weak. The interaction energy between a vibration system with the weak coupling and a vibration system with the strong coupling to the heat bath is given in Fig. 11.6 too (the dashed and dotted curve denoted by "weak-strong"). In this case, the absolute values of the interaction energy are more that about one order greater than the corresponding values in the system with the strong coupling to the heat bath but considerably smaller than the values in the system with the weak coupling to the heat bath.

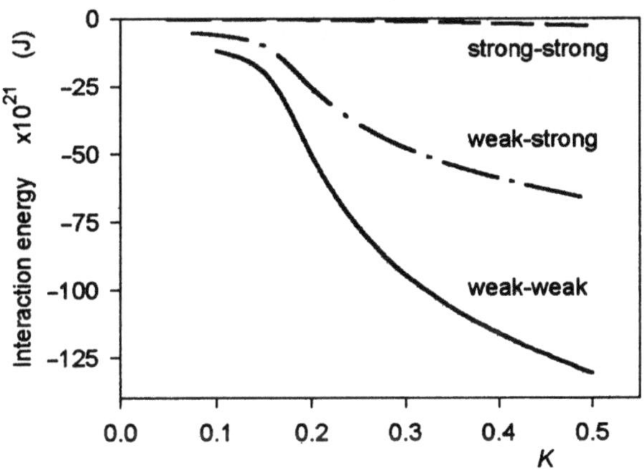

Fig. 11.6 The interaction energy I of two joint interacting vibration systems as a function of the interaction coefficient K. The transition probabilities Φ_i, χ_{ij} and the frequencies of normal modes before interaction are the same as in Fig. 11.5. $S = 3\,\mathrm{s}^{-1}$, $n_{\max} = 3500$.

Similar dependence of the interaction energy on the coupling to the heat bath (i.e. on χ) can be derived for various values of $n_{\max}$. The smaller the $n_{\max}$ value the smaller the absolute value of the interaction energy. For $n_{\max} = 300$ the absolute values of the interaction energy are about one order smaller than the values given in Fig. 11.6. Nevertheless, the typical dependence on the coupling to the heat bath remains preserved. For the strong coupling to the heat bath the interaction energy is much smaller than for the weak coupling.

We assumed that only the resonant interaction between vibration systems exists, i.e. that the vibration modes with the same frequencies interact with one another. The non-resonant interactions between the vibration modes with different frequencies are neglected. Interactions between all spectral lines cause splitting up the frequencies of the vibration modes into a complicated spectral pattern. The splitting up (i.e. the frequencies of the spectral lines) depends on the interaction between the systems.

The interaction between the vibration systems depends on the intensity of the generated electromagnetic field of particular spectral components and, therefore, on the energy condensed in particular vibration modes, on the distance between the vibration systems, and on the permittivity and conductivity of the surrounding medium. The frequencies of the vibration modes depend on the condensed energy.

The intensity of the electric field mediating interactions depends on the permittivity and conductivity of the medium around the vibration systems. The main component of the surrounding medium is water. A strong absorption of electromagnetic waves occurs in the region of anomalous dispersion and the interaction coefficient may be very small. Anomalous dispersion is in the regions of the bending and stretching vibrations of water molecules. As a result the frequencies of the vibration modes in this region remain unchanged. This may have a strong effect on the interaction energy.

12. Coherent States in Cancer Cells

12.1 Phosphorylation of Proteins

Protein functions are regulated through selective phosphorylation of certain amino acids. Protein phosphorylation is the most ubiquitous intracellular control mechanism regulating certain metabolic pathways, gene transcription and translation, membrane transport, hormonal response, muscle contraction, photosynthesis, cell division and cell growth, viral oncogene response, and memory [163–165]. Phosphorylation changes the side chain residues of certain amino acids in proteins. The terminal phosphate group (inorganic phosphate) of ATP is transferred to a target protein. The phosphate group is attached to some amino acid side chain residues containing the hydroxyl group OH where phosphorus with three oxygen atoms and two hydrogen atoms is substituted for hydrogen. In eukaryotic cells, reversible phosphorylation catalysed by protein kinases (and the opposite process by phosphatases) occurs on the serine, threonine and tyrosine residues. Phosphorylation of a histone H4 protein by histidine kinase [165] was reported too. Phosphorylation in healthy cells is found predominatly on the residues of serine and threonine. Phosphorylation in cancer cells is different. For the sake of further explanation phosphorylation of the side chain residues of serine, threonine, and tyrosine is shown in Fig. 12.1 [166]. The arrows symbolically indicate replacement of the hydrogen atom in the OH group by the phosphate group.

Phosphorylation may promote various changes in proteins which may be found e.g. in [165]. Generally, protein phosphorylation can result in stimulation or inhibition of enzyme activity, promote local or remote conformation changes, alter the protein surface properties, and affect self-association with or recognition by other proteins. In many cases the major incentive for conformation changes appears to be electrostatic.

Local conformation changes may be associated with long range effects leading to activation of catalytic function, e.g. in glycogen phosphorylase, the mechanism of activation by phosphorylation is allosteric – local conformation changes are associated with long range effects so that phosphorylation at Ser14 leads to activation at the catalytic site over 3.5 nm away. (Conformation changes which couple two widely separated binding sites of a molecule are called allosteric transitions. The allosteric mechanism makes possible to link the fate of one molecule to the presence or absence of any other.) Mechanism based on electrostatic blocking processes may change functions of proteins (e.g. in isocitrate dehydrogenase phosphorylation of Ser113 prevents binding of the anionic substrate without promoting any conformational change). The effects of phosphorylation may appear to be mostly conformational, e.g. the assumption is that in cyclic AMP (adenosine monophosphate) dependent protein kinase denoted as PKA phosphorylation of Thr197 localizes a loop that forms a part of the substrate recognition site and the regulatory-subunit

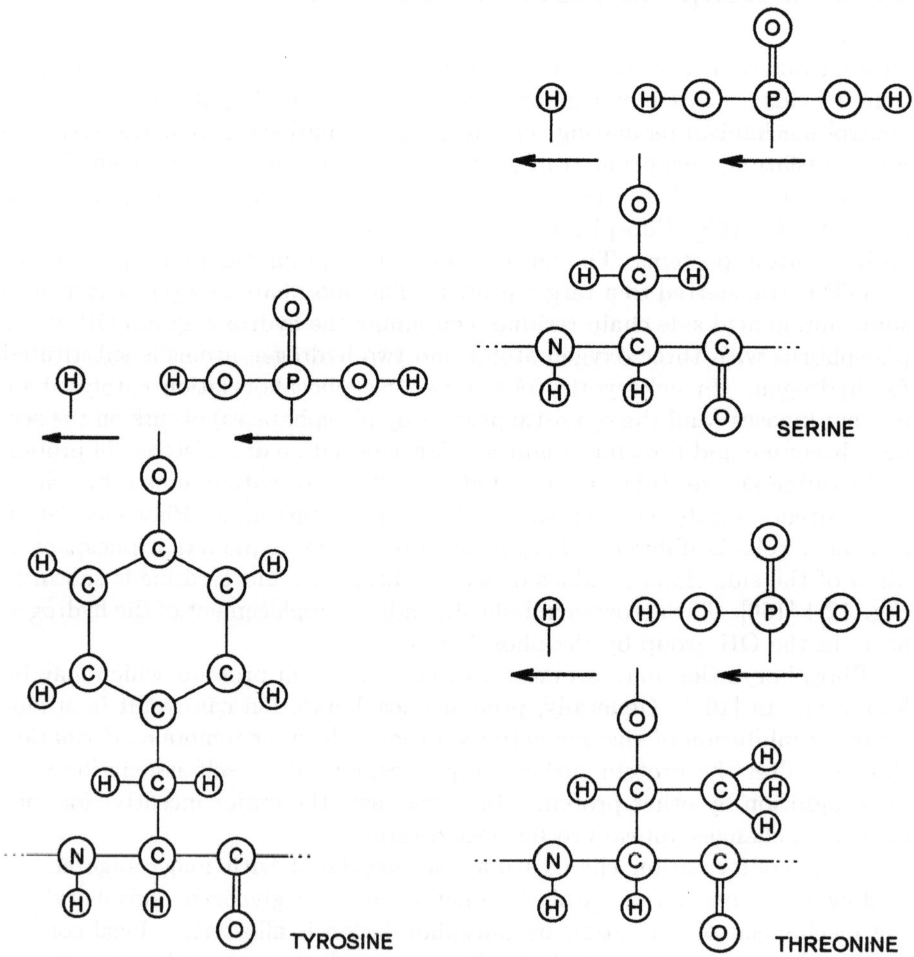

Fig. 12.1 A schematic picture of phosphorylation of the side chain residues of serine, threonine, and of tyrosine. Hydrogen atoms are replaced by phosphate groups.

recognition site and maintains the catalytic-site residues in their correct conformations. Protein–protein interaction is inhibited through changes in the surface properties and conformation so that protein is not recognized by the enzyme (e.g. in the protein HPr by phosphorylation of Ser46). Effect of phosphorylation on the catalytic sites and the ATP binding sites of kinases is described e.g. in [165], too. Electric microdipoles may be arranged by alignment of the peptide dipoles parallel to the protein helix axis. The universal control mechanism of the cell cycle depends on phosporylation of p34 kinase which forms complexes with different types of proteins called cyclins [163,164]. The connection between cyclins as the cell cycle regulators and cancer transformation is given in [167]. The effects described above give evidence for regulation by a single phosphate site. An important feature of control by phosporylation is multiple phosphorylation in which several sites may be phosphorylated or dephosphorylated in response to different enzymes. At present we have no evidence for the structural basis of the mechanism of multiple phosphorylation.

Phosphorylation of proteins may have no visible effects at all (silent phosphorylation) [165]. For example, pepsinogen and pepsin are phosphorylated at Ser68. Ezymatic removal of phosphate has no effect on activity of either pepsinogen or pepsin. The function of the phosphate is unknown. The role of phosphate in ovalbumin is not known either but one phosphorylation site is close to the analogous reaction centre of the serine protease inhibitors (serpins) [165]. The silent phosphorylation makes no defined changes in proteins and suggests that as yet unidentified interactions between proteins may exist [165].

12.2 Cancer Process

Genes encode proteins that provide the structure of living cells and catalyse the biochemical reactions in cells. Proto-oncogenes are benign genes in normal cells. Altered versions of proto-oncogenes are oncogenes, which are linked with cancer process. Several mechanisms may cause the alteration of proto-oncogenes. It may well be that creation of oncogenes is a consequence of point mutation or specific chromosomal translocation. Whatever the specific mechanism transforming a proto-oncogen into oncogene may be an oncogen must have its final effect by way of the proteins encoded by it. Proteins encoded by the oncogenes differ from the normal proteins. For instance, the Rous sarcoma virus gene [168,169], which is transcribed "backward" into DNA when the virus infects a chicken cell, encodes the protein p60*src*. The p60*src* is a protein kinase; enzyme, which catalyses phosphorylation – addition of a phosphate group to other proteins.

As was already mention phosphorylation is a general controling mechanism in living cells regulating protein functions. The p60*src* protein kinase in virus infected chicken cells causes phosphorylation of the residues of tyrosine. Even normal cells have certain amount of phosphotyrosines in their proteins but except as a transient phenomenon phosphorylation of the residues of tyrosine is a rare event in normal cells. The majority of oncogenes (e.g. *yes*, *abl*, *src*, etc.) encodes tyrosine protein kinases (with the exception of *ras* oncogene – p21*ras* protein kinase catalyses on threonine). Evidence on tyrosine phosphorylation in human cancer cells is given e.g. in [170]. The target proteins that are phosphorylated by protein kinases are e.g. vinculin, histone, p36, and others.

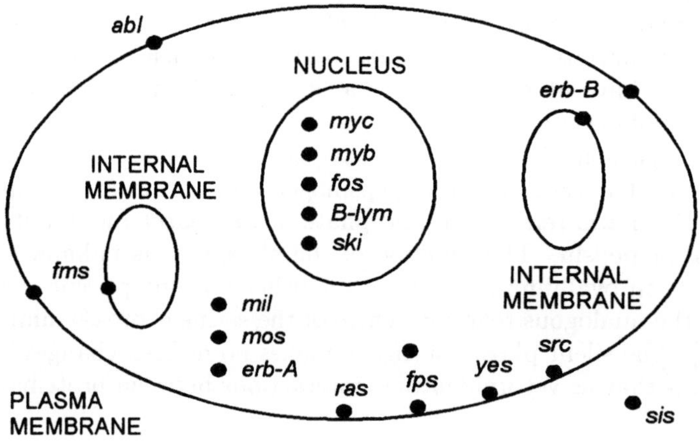

Fig. 12.2 Sites of action of some oncogene products in cells. A schematic picture.

Sites of action of some oncogene products in cells are shown in Fig. 12.2. For instance, vinculin may be found in adhesion plaques, small patches on the bottom of a cell adhering closely to the surface on which it grows. Cytoskeletal filaments containing the protein actin terminate at adhesion plaques. (Actin filaments are two stranded helical polymers with a diameter 5–9 nm. They are organized in a variety of linear bundles, two dimensional networks, and three dimensional gels. These flexible structures are mostly highly concentrated in the cortex just beneath the plasma membrane. Actin filaments, microtubules, and intermediate filaments form a complex network – the cytoskeleton – that extends throughout the cell interior.) Vinculin is interposed between the plasma membrane and the ends of the actin filament bundles and may play a role of the mediator connecting the bundles to the membrane. Nevertheless, our present knowledge of the function of phosphorylated vinculin as well as of other phosphorylated proteins is still considerably limited.

Most cancers are probably initiated by a change in the cell's DNA sequence leading to the change of proto-oncogenes to oncogenes. Therefore, mechanism of oncogenes creation and expression is widely studied. The final effects of oncogenes is carried out by the proteins encoded by them. We will not discusse here the pathway from the phosphorylated target proteins to creation of a malignant cell. First of all we will describe the behaviour of malignant cells. The basic properties of cancer cells and their behaviour were well known to classical oncologists half a century ago. The deviations of malignant cells from normal cells were summarised in the following categories [171]:

a) Local invasion.

The malignant cells at certain stage of tumor development insinuate themselves amongst the surrounding normal cells moving outwards from the locus of their origin. The cell surface and the surrounding structures with which the cell surface comes into contacts are of profound importance for the cell invasive movement.

b) Metastasis.

The malignant invasive cells reach moving body fluids, separate themselves and are passively transported to all parts of the body. They become widely disseminated. Adhesion and binding of tumor cells is very often organ specific [172]. The passive spread of cancer cells and the consequent establishment of new foci of growth and invasion, probably involve the properties of the malignant cell surface. It can hardly be doubted that adhesiveness plays a major part in determining the initial detachment of the malignant cell, and it may also play a part in their arrest in other parts of the body.

c) Disorganisation.

The malignant cells fail to form tissue of the arrangement characteristic of their normal counterparts. It should be pointed out that simplification of the shape and the disorganization of the arrangement may be at least in part an expression of surface abnormalities of the malignant cells.

d) Proliferation.

The malignant cells keep up a persistent growth by mitosis. There is some possibility that the growth activity of malignant cells may be linked with persistent surface activity associated with diminished adhesiveness.

Invasion and metastasis were considered by the classical oncologists to be crucial aspects of malignancies [171] and referred to the surface properties of malignant cells. Contemporary oncological research confirms the classical assumptions. Molecular biology elucidates the mechanisms of adhesion and the role of the cell surface. Cell–cell adhesion and cell–substrate adhesion are mediated by two major categories of molecules, the cell adhesion molecules (CAM) and the cell–substrate adhesion molecules (SAM), respectively. The adhesion mechanisms are important for biological function of cells [173]:

Cell–cell and cell–matrix interactions are tightly regulated processes which can in turn regulate cell shape, determine cell polarity and tissue architecture, trigger cell motility, and induce the expression of receptors for growth factors thereby affecting cellular proliferation.

There are four classes of adhesion receptors on the cell surface: cadherins, integrins, selectins, and members of the immunoglobulin (Ig) superfamily [1,173]. Cadherins are Ca^{2+} dependent transmembrane molecules primarily responsible for the tissue specific cell–cell adhesion. Classical cadherins have five extracellular domains – three of which contain Ca^{2+} binding sites. The total length of the extracellular chain is of about 20–30 nm. The cytoplasmic part interact with actin filaments of the cytoskeleton via catenin proteins (α-, β-, and γ-catenins). The cadherins cannot promote cell adhesion unless they are complexed with the catenins. Phosphorylation of β-catenin leads to decrease in the strength of cell adhesion. The loss of E-cadherins mediated cell adhesion is currently thought to initiate the transformation of epithelial cells into invasive cells.

Integrins act as a link in the chain of adhesive interactions between cells and their extracellular microenvironment (except for $\beta 2$ integrins mediating cell–cell adhesion). Integrins are "heterodimers" containing two chains: α- and β-chains. β_1 and β_3 integrin families are crucially important because they are the main way for cells both bind and respond to the extracellular matrix. The integrin molecule has a globular binding head about 25 nm apart from the plasma membrane bilayer. The protein serves as a transmembrane linker binding to the actin cytoskeleton filaments via attachment proteins talin and α-actin. Integrins must interact with the cytoskeleton in order to bind cells to extracellular matrix. A transmembrane attachment to the cytoskeleton appears to be a fundamental requirement for both cell–matrix and cell–cell adhesions. Integrins mediate communication between the extracellular matrix and the cytoskeleton which can influence the organization of cell's cytoskeleton.

The immunoglobulin (Ig) superfamily of proteins is responsible for Ca^{2+} independent cell–cell adhesion. The molecules contain one or more Ig-like domains that are characteristic of antibody molecules. The extracellular part of the polypeptide chain is folded into five immunoglobulin-like domains. Most of them are single-pass transmembrane proteins which are involved in the cell signaling or binding to the cytoskeleton. The adhesions mediated by members from the immunoglobulin superfamily of proteins are much weaker than adhesions of cadherins. The proteins from the Ig superfamily very likely contribute more to regulation or fine tuning of the adhesive interactions. Evidence has accumulated supporting a role of Ig superfamily members in initiating signal transduction cascades that lead to gene regulation and cell differentiation. Exactly how signals are transmitted from the extracellular to the cytoplasmic side of the cell is unclear. It is assumed that tyrosine phosphorylation

– dephosphorylation is involved in the signaling pathway as a fundamental mechanism.

Selectin adhesion molecules mediate special interaction of the venular wall and leukocytes. They are involved in leukocyte recruitment to sites of both acute and chronic inflammation, and may be involved in a wide variety of human diseases including cancer metastasis.

The role of adhesion receptors in detachment of individual cells and in invasion process is discused e.g. in [173]:

It has become clear that the process of malignant progression is accompanied from its onset by profound changes in the expression and function of cell adhesion molecules. Malignant transformation may trigger the expression of new receptors, result in the loss or suppression of others, or lead to expression of genetic variants of receptors expressed on normal cells. These changes eventually give rise to cells with modified repertoires of adhesion receptors whose potential to communicate with the microenvironment has been irreversibly altered. These alterations are at the root of the cancer cell's potential to dissociate from its neighboring cells, migrate through the stroma, invade and exit blood vessel, anchor in a foreign microenvironment and respond to growth modulating cytokines in a modified manner – processes which are the hallmark of cancer progression and metastasis.

Initiation of an invasive metastatic process is signalled by the loss of junction contact between adjacent cells in the cancerous tissue. This process is accompanied by a loss of the differentiated morphology. The loss of tissue cohesiveness which is essential for initiation of tumor cell migration involves changes in the expression and function of several mediators of cellular communication. Alteration in E-cadherin expression and function constitute a major underlying mechanism in the acquisition of invasive properties by cells of epithelial origin. Nevertheless, E-cadherin function may be altered by mechanism other than the direct quantitative reduction in cell surface expression.

The disruption of cellular junctions cannot by itself account for the increased migration and invasive capacity. The disruption of cell–extracellular matrix adhesion is necessary too. Nevertheless, the data on integrin expression in transformed and malignant cells defies generalization and is cell system- and integrin-specific. The findings on $\alpha_2\beta_1$-integrins serve to reinforce the conclusion that the contribution of particular integrins to malignant phenotype may ultimately be dictated by cellular context in which they are expressed. The cells may be released from their dependence on integrin mediated signaling for essential cellular function such as growth and migration.

The adhesion changes of tumor cells are summarized in the following way [173]:

In general, increased homotypic adhesion of tumor cells to other tumor cells in the primary tumor site appears to prevent cellular exit from the primary tumor and thus, inhibits metastasis. In contrast, tumor cell adhesion to endothelial cells or subendothelial matrix components represents important early steps of the extravasation process and consequently, serve to promote metastasis.

First of all we have to point out that adhesion function of both cadherins and integrins is conditioned by their binding to the cytoskeleton, i.e. the cytoskeleton together with these transmembrane molecules seems to form a common functional system. The cytoskeleton has an important function in cell adhesion. The experimental data suggest, that very often functional alterations are more important than quantitative expression changes. Striking disorder in the system of invasive cells may be an important feature too. Regardless of the effect of phosphorylated catenins on adhesion mediated by cadherins the total effect of phosporylated proteins and their general function in a cancer cell are still rather unclear. The link between phosphorylated proteins and malignant cells behaviour is still not satisfactorily investigated. A schematic picture of the chain of successive steps of malignant transformation of a cell is given in Fig. 12.3.

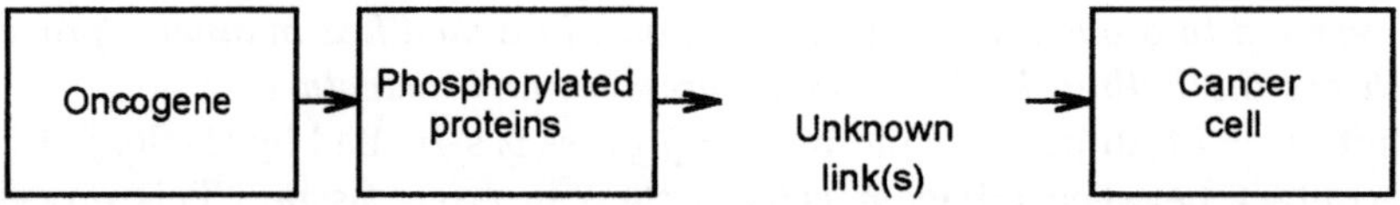

Fig. 12.3 A schematic picture of the chain of malignant transformation of a cell corresponding to present knowledge.

Malignant cells have tendency to secret a large amount of lactic acid as a result of the increased breakdown of glucose by glycolysis, an anaerobic pathway of producing energy to cover their large energy consumption. We may ask a question – to which extent the increased energy consumption may be connected with the not yet identified link of malignant transformation between phosphorylated proteins and changed cell surface properties.

12.3 The Role of Protein Phosphorylation in Cancer Mechanism – Hypothesis

We will discuss the mechanism of phosphorylation considering the coherent vibration states in proteins. First of all we have to mention some properties of the phosphate group. The electronegativities of phosphorus, hydrogen, and oxygen are 2.1, 2.1, and 3.5, respectively. The phosphorus and hydrogen atoms have positive charges and the oxygen atoms have negative charges. The phospate group is an electric multipole. Therefore, after phosphorylation the charge distribution in the side chain residue is altered. Besides the altered charge distribution the phosphate group represents a large mass. The mass of the phosphate group is about 80 times greater than the mass of the hydrogen atom in the OH group which is replaced by the phosphate group.

The side chain residue charges have a dominant effect on the charge distribution in proteins [119]. The charge distribution in a phosphorylated protein differs from the charge distribution in a protein before phosphorylation. The

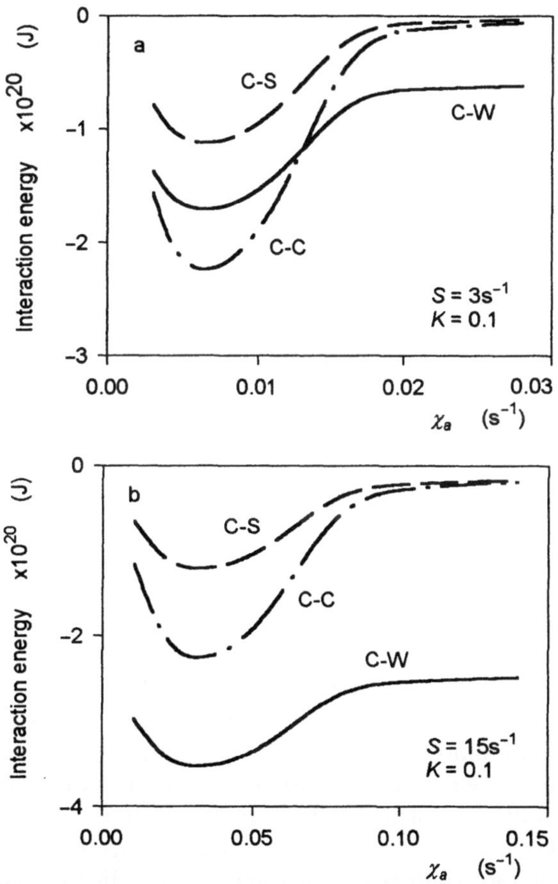

Fig. 12.4 Interaction energy of model systems as a function of the non-linear transition probability χ_a ($\chi_b = 0.1\chi_a$, $\Phi = 10\chi_a$, $K = 0.1$). The energy supply is small – $S = 3\,\mathrm{s}^{-1}$ (a) and great – $S = 15\ \mathrm{s}^{-1}$ (b). C-S: interaction between a system with current value of χ_a (given on the abscissa) and a system with the strong coupling to the heat bath ($\chi_a = 0.2\,\mathrm{s}^{-1}$). C-W: interaction between a system with current value of χ_a and a system with the weak coupling to the heat bath ($\chi_a = 0.001\,\mathrm{s}^{-1}$). C-C: interaction between two systems with current values of χ_a.

electric field in the protein is changed and, therefore, the properties of the vibration system too. Even phosphorylation in one point (as well as phosphorylation in several discrete points) may have significant effects on the vibration systems in proteins. Massive phosphorylation has to alter the vibrations in a protein completely. The elastic force constants in a protein may be considerably influenced which results in frequency alterations.

Charges in the phosphorylated side chain residues interact with the protein ambient medium, i.e. with the heat bath. A strong Coulomb coupling to the heat bath occurs. Increased coupling to the heat bath causes an increase of the linear and the non-linear energy transfer between the vibration system and the heat bath. The effect of the coupling to the heat bath on energy condensation in coherent vibration states is shown in Figs 11.1–11.5. A strong coupling reduces the energy condensed in the vibration system. The electromagnetic field generated by the protein and by the structures of proteins depends on the energy condensed in their vibration systems. The electromagnetic field mediates interaction between vibration systems in proteins. Therefore, interaction between proteins may depend on phosphorylation.

Due to interactions vibrations in particular proteins and protein structures may form a coherent vibration system in a cell. The cell vibration system generates an electromagnetic field whose intensity depends on phosphorylation of proteins. The generated electromagnetic field may be one of the factors determining the surface properties of a cell as well as interaction with other cells. Healthy cells are highly excited, i.e. energy condensed in the vibration modes is high. The generated electromagnetic field has very likely some spectral components with high intensity of the electric field. After malignant transformation phosphorylation causes a strong coupling to the protein surroundings and consequently great energy leakage from the normal modes. The energy sources of a transformed cell are very likely not sufficient to maintain the normal excitation level. Total energy condensed in the vibration system in a malignant cell is smaller than energy condensed in the vibration sytem in a normal cell and depends on the level of phosporylation, i.e. on the number of phosphorylated amino acid residues in proteins, as well as on the level of energy supply from the metabolic sources. The intensity of the generated electromagnetic field in transformed cells is smaller than the intensity in a normal cell and especially the high frequency components are very likely suppressed.

Long range attractive forces among malignant cells mediated by the oscillating electric field generated by the coherent vibrations are very likely weak if the coupling to the heat bath is strong. Fig. 11.6 shows the interaction energy between systems with the strong coupling to the heat bath (the curve denoted by "strong-strong"). We analyzed interaction energy of two vibration systems as a function of the non-linear coupling to the heat bath. Fig. 12.4 shows interaction energy as a function of χ for a model system with the same normal modes as in the model shown in Fig. 11.1. The energy supply is small ($S = 3\,s^{-1}$ – Fig. 12.4a) and great ($S = 15\,s^{-1}$ – Fig. 12.4b). The interaction energy depends on the coupling of the vibration system to the heat bath. The curves denoted by C-C, C-S, and C-W may be related to two identical malignant cells of a tumour in different phases of its development, to a malignant cell at the boundary of a tumour and a malignant cell inside the tumour, and to a tumour cell and its healthy counterpart, respectively. If the level of phosphorylation in the tumour cells is low (very likely in the initial phases of tumour development) the interaction energy is great and the tumour cells are kept together (C-C curve for $\chi_a \leq 0.01\,s^{-1}$ and for $\chi_a \leq 0.05\,s^{-1}$ in Fig. 12.4a and Fig. 12.4b, respectively). If the level of phosphorylation is greater than corresponds to $\chi_a = 0.015\,s^{-1}$ (Fig. 12.4a) or to $\chi_a = 0.07\,s^{-1}$ (Fig. 12.4b) the interaction energy between malignant cells (the C-S and the C-C curves) is considerably smaller than the interaction energy between a malignant cell and its healthy counterpart (the C-W curve). The self-consistent field of normal cells in healthy tissue surrounding the region occupied by a tumour acts

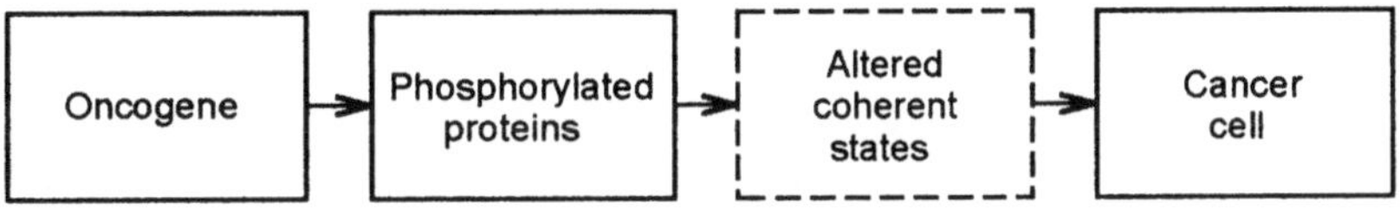

Fig. 12.5 A schematic picture of the chain of malignant transformation of a cell including alteration of coherent states.

on the malignant cells by attractive forces. This may correspond to a model in which the malignant cells are "pulled out" from the tumour region inside the healty cells. Invasive growth of the malignant cells may include the electromagnetic interaction with the normal cells. Very weak forces may act among the malignant cells inside a healthy tissue and the malignant cells in the region of tumour. The attractive forces between the healthy cells and the malignant cells are weaker than the forces between the healthy cells (Fig. 12.4).

Fig. 12.5 shows a hypothetical role of coherent vibrations in malignant transformation of a cell.

We assume that the mechanism of the coherent electromagnetic field may act – to a certain degree – in parallel with chemical adhesion system. This electromagnetic mechanism mediates long range interaction beyond the distance of adhesion of cadherins, integrins, proteins from immunoglobulin superfamily, and selectins and can have a profound role in organization of adhesion.

The frequencies of particular modes and, therefore, the frequencies of the spectral components of the electromagnetic field may be changed by phosphorylation. The potential profile of oscillators in phosphorylated protein molecules may differ from the potential profile in normal protein molecules.

Some oscillators may be non-linear and the frequencies of oscillators may depend on the energy level. We may expect that the spectral components of the electromagnetic field of malignant cells are different from the spectral components of normal cells.

Hypothesis presented in this section was not verified experimentally yet. We have to emphasize here that invasiveness and metastatic growth of a cancer cell include many other factors of chemical, biophysical, and biological nature. Any biological event is a result of a set of elementary processes of various kinds taking place simultaneously and successively, mutually bound and influencing one another. Biophysical processes are responsible for creation and maintenance of order.

These elementary processes are important tools for ordering and self-assembly of biological systems. As far as we do not know them, their mutal dependence, and time succession we cannot decipher not only the enigma of cancer but we do not understand the life either.

13. Appendices

13.1 Heisenberg, Schrödinger, and Interaction Representations

We describe three customary ways of introducing the time development of a system in quantum mechanics [174]. The time dependent Schrödinger equation is given in the form

$$H\,\Psi_{\mathrm{S}}(t) = \mathrm{i}\hbar \frac{\partial}{\partial t} \Psi_{\mathrm{S}}(t) \tag{13.1}$$

where $\Psi_{\mathrm{S}}(t)$ is the time dependent wave function and H is the time independent Hamiltonian. The usual form of the Schrödinger eigenvalue equation at the time $t = 0$ is $H\,\Psi_{\mathrm{S}}(0) = E\,\Psi_{\mathrm{S}}(0)$ where E is the eigenvalue of energy. The formal solution of Eq. (13.1) is given in the form

$$\Psi_{\mathrm{S}}(t) = \exp\left(-\frac{\mathrm{i}}{\hbar}Ht\right)\Psi_{\mathrm{S}}(0) = \exp\left(-\frac{\mathrm{i}}{\hbar}Et\right)\Psi_{\mathrm{S}}(0) \tag{13.2}$$

where $\Psi_{\mathrm{S}}(0)$ is the wave function at $t = 0$.

In the Schrödinger representation operators do not depend on time explicitly and the wave functions depend on time.

In the Heisenberg representation the wave function Ψ_{H} is defined by the relation

$$\Psi_{\mathrm{H}} = \Psi_{\mathrm{S}}(0) \tag{13.3}$$

and, therefore,

$$\Psi_{\mathrm{H}} = \exp\frac{\mathrm{i}}{\hbar}Ht\,\Psi_{\mathrm{S}}(t)\,. \tag{13.4}$$

The Heisenberg operator A_{H} corresponding to the operator A_{S} in the Schrödinger representation is determined by the unitary transformation

$$A_{\mathrm{H}} = \exp\left(\frac{\mathrm{i}}{\hbar}Ht\right)A_{\mathrm{S}}\exp\left(-\frac{\mathrm{i}}{\hbar}Ht\right)\,. \tag{13.5}$$

A unitary transformation of operators and wave functions makes possible transition from one description of the quantum mechanical problem to another one. The unitary transformation does not alter the matrix element of any operator. All relevant physical properties and parameters are determined by the matrix elements.

In the Heisenberg representation operators depend on time explicitly and the wave functions do not depend on time. The equation of motion in the Heisenberg representation may be determined from Eq. (13.5) by differentiation. One gets

$$\begin{aligned}\frac{\partial A_{\mathrm{H}}}{\partial t} &= \frac{\mathrm{i}}{\hbar} H A_{\mathrm{H}} + \exp\left(\frac{\mathrm{i}}{\hbar}Ht\right) A_{\mathrm{S}} \left[-\frac{\mathrm{i}}{\hbar} H \exp\left(-\frac{\mathrm{i}}{\hbar}Ht\right)\right] \\ &= \frac{\mathrm{i}}{\hbar}\left(H A_{\mathrm{H}} - A_{\mathrm{H}} H\right) . \end{aligned} \tag{13.6}$$

Therefore, we may write the equation of motion in the form

$$\frac{\partial A_{\mathrm{H}}}{\partial t} = -\frac{\mathrm{i}}{\hbar}\left[A_{\mathrm{H}}, H\right] . \tag{13.7}$$

The Hamiltonian H has the same form in the Schrödinger and in the Heisenberg representation as follows from the relation

$$H_{\mathrm{H}} = \exp\left(\frac{\mathrm{i}}{\hbar}Ht\right) H \exp\left(-\frac{\mathrm{i}}{\hbar}Ht\right) = H .$$

The Interaction representation may be used for a system with the time dependent perturbation. The Hamiltonian of the system in question is assumed to be in the form

$$H = H_0 + H'(t) \tag{13.8}$$

where H_0 does not depend on time and the perturbation Hamiltonian H' may depend on time explicitly. The Schrödinger wave function and the Schrödinger operators are transformed to the interaction form by the relations

$$\Psi_{\mathrm{I}}(t) = \exp\left(\frac{\mathrm{i}}{\hbar}H_0 t\right) \Psi_{\mathrm{S}}(t) \tag{13.9}$$

$$A_{\mathrm{I}} = \exp\left(\frac{\mathrm{i}}{\hbar}H_0 t\right) A_{\mathrm{S}} \exp\left(-\frac{\mathrm{i}}{\hbar}H_0 t\right) . \tag{13.10}$$

The interaction Hamiltonian is given by the relation

$$H_{\mathrm{I}} = \exp\left(\frac{\mathrm{i}}{\hbar}H_0 t\right) H' \exp\left(-\frac{\mathrm{i}}{\hbar}H_0 t\right) . \tag{13.11}$$

The interaction operators as well as the interaction wave functions depend on time explicitly.

We will derive the equation of motion in the interaction representation. We will differentiate Eq. (13.9) and use the time dependent Schrödinger equation

$$\mathrm{i}\hbar \frac{\partial \Psi_{\mathrm{S}}(t)}{\partial t} = (H_0 + H')\, \Psi_{\mathrm{S}}(t) \, .$$

We get

$$\begin{aligned}
\frac{\partial \Psi_{\mathrm{I}}}{\partial t} &= \frac{\mathrm{i}}{\hbar} H_0 \, \Psi_{\mathrm{I}} + \exp\left(\frac{\mathrm{i}}{\hbar} H_0 t\right) \frac{\partial \Psi_{\mathrm{S}}}{\partial t} \\
&= \frac{\mathrm{i}}{\hbar} H_0 \, \Psi_{\mathrm{I}} - \frac{\mathrm{i}}{\hbar} \exp\left(\frac{i}{\hbar} H_0 t\right) (H_0 + H')\, \Psi_{\mathrm{S}} \\
&= -\frac{\mathrm{i}}{\hbar} \exp\left(\frac{\mathrm{i}}{\hbar} H_0 t\right) H' \exp\left(-\frac{\mathrm{i}}{\hbar} H_0 t\right) \Psi_{\mathrm{I}}(t) \, . \qquad (13.12)
\end{aligned}$$

Therefore, the equation of motion in the interaction representation is given in the form

$$\mathrm{i}\hbar \frac{\partial \Psi_{\mathrm{I}}}{\partial t} = H_{\mathrm{I}} \, \Psi_{\mathrm{I}} \, . \qquad (13.13)$$

Equation (13.13) is similar to the ordinary time dependent Schrödinger equation. Nonetheless, the Hamiltonian H_{I} is derived from the perturbation Hamiltonian H'.

13.2 Commutation Relations

a) For evaluation of operators in the interaction representation we need the relations

$$\begin{aligned}
a_j \exp\left(-\frac{\mathrm{i}}{\hbar} H_0 t\right) &= \exp\left(-\frac{\mathrm{i}}{\hbar} H_0 t\right) \exp\left(-\mathrm{i}\omega_j t\right) a_j \\
a_j^+ \exp\left(-\frac{\mathrm{i}}{\hbar} H_0 t\right) &= \exp\left(-\frac{\mathrm{i}}{\hbar} H_0 t\right) \exp\left(\mathrm{i}\omega_j t\right) a_j^+ \, . \qquad (13.14)
\end{aligned}$$

Equation (13.14) may be easily derived if one expands the exponential operator function $\exp\left(-\mathrm{i}H_0 t/\hbar\right)$ into a power series

$$\begin{aligned}
\exp\left(-\frac{\mathrm{i}}{\hbar} H_0 t\right) &= 1 + \left(-\frac{\mathrm{i}}{\hbar} H_0 t\right) + \frac{1}{2!}\left(-\frac{\mathrm{i}}{\hbar} H_0 t\right)^2 + \frac{1}{3!}\left(-\frac{\mathrm{i}}{\hbar} H_0 t\right)^3 + \ldots \\
&\quad + \frac{1}{n!}\left(-\frac{\mathrm{i}}{\hbar} H_0 t\right)^n + \ldots \qquad (13.15)
\end{aligned}$$

where the Hamiltonian H_0 is given by the relation

$$H_0 = \sum_i \hbar\omega_i a_i^+ a_i \; . \tag{13.16}$$

We have to interchange the position of operators in the terms

$$\begin{aligned} a_j \left[\sum_i \omega_i a_i^+ a_i\right]^n &= \left[\sum_i \omega_i \left(a_i^+ a_i a_j + \delta_{ij} a_i\right)\right]\left[\sum_i \omega_i a_i^+ a_i\right]^{n-1} \\ &= \left[\sum_i \omega_i \left(a_i^+ a_i + \delta_{ij}\right)\right] a_j \left[\sum_i \omega_i a_i^+ a_i\right]^{n-1} \\ &= \left[\sum_i \omega_i \left(a_i^+ a_i + \delta_{ij}\right)\right]^2 a_j \left[\sum_i \omega_i a_i^+ a_i\right]^{n-2} \\ &= \left[\left(\sum_i \omega_i a_i^+ a_i\right) + \omega_j\right]^n a_j \end{aligned} \tag{13.17}$$

where the commutation relations $[a_j,\, a_i^+] = \delta_{ij}$ and $[a_j,\, a_i] = 0$ for the operators a_i^+, a_i and a_j were used. After substitution of Eq. (13.17) into Eq. (13.15) (for $n = 1, 2, 3, \ldots$, i.e. for all the terms of the series) and after replacing the series by the exponential function we get the first relation in (13.14). Likewise, we may determine the second relation in (13.14).

b) Eq. (13.14) enable us to determine the relations

$$\begin{aligned} (a_j)^m \exp\left(-\frac{\mathrm{i}}{\hbar} H_0 t\right) &= (a_j)^{m-1} \exp\left(-\frac{\mathrm{i}}{\hbar} H_0 t\right) \exp(-\mathrm{i}\omega_j t) a_j \\ &= \exp\left(-\frac{\mathrm{i}}{\hbar} H_0 t\right) \exp(-\mathrm{i}m\omega_j t)(a_j)^m \\ (a_j^+)^m \exp\left(-\frac{\mathrm{i}}{\hbar} H_0 t\right) &= \exp\left(-\frac{\mathrm{i}}{\hbar} H_0 t\right) \exp(\mathrm{i}m\omega_j t)(a_j^+)^m \end{aligned} \tag{13.18}$$

where H_0 is given by Eq. (13.16).

c) For evaluation of the Hamiltonian describing the multiple quanta processes we need the relations

$$\begin{aligned} (a_j)^m a_i^+ &= (a_j)^{m-1}(a_i^+ a_j + \delta_{ij}) \\ &= (a_j)^{m-1} a_i^+ a_j + \delta_{ij}(a_j)^{m-1} \\ &= (a_j)^{m-2}(a_i^+ a_j + \delta_{ij}) a_j + \delta_{ij}(a_j)^{m-1} \\ &= a_i^+ (a_j)^m + m\delta_{ij}(a_j)^{m-1} \\ (a_j^+)^m a_i &= a_i (a_j^+)^m - m\delta_{ij}(a_j^+)^{m-1} \end{aligned} \tag{13.19}$$

where $(a_j)^m$ and $(a_j^+)^m$ are $a_j a_j a_j \dots a_j$ m-times and $a_j^+ a_j^+ a_j^+ \dots a_j^+$ m-times, respectively.

d) With the aid of Eq. (13.19) we determine the relations

$$\begin{aligned}
(a_i^+)^m (a_i)^m &= a_i^+ a_i (a_i^+)^{m-1} (a_i)^{m-1} - (m-1)(a_i^+)^{m-1} (a_i)^{m-1} \\
&= \left[a_i^+ a_i - (m-1)\right] (a_i^+)^{m-1} (a_i)^{m-1} \\
&= \left[a_i^+ a_i - (m-1)\right] \left[a_i^+ a_i - (m-2)\right] \dots \left[a_i^+ a_i - 1\right] a_i^+ a_i \\
(a_i)^m (a_i^+)^m &= \left[a_i^+ a_i + m\right] \left[a_i^+ a_i + (m-1)\right] \left[a_i^+ a_i + (m-2)\right] \dots \left[a_i^+ a_i + 1\right]
\end{aligned} \tag{13.20}$$

which one needs to evaluate expectation values of the multiple quanta processes.

13.3 The Adiabatic Hypothesis and Operator of Dynamic Evolution

We assume that an unperturbed system described by the Hamiltonian H_0 may be transformed into a perturbed system described by the Hamiltonian $H_0 + H_1$ (where neither H_0 nor H_1 depend on time) by gradually switching on of a small perturbation given by the Hamiltonian H_1. We will use the perturbation term in the form $H_1 \exp(st)$ where $s > 0$ is a small number (we assume that $\exp(st) \approx 1$ for a reasonably great t). For $t \to -\infty$ this operator equals zero and the unperturbed system is described by the Hamiltonian H_0 and by the wave vectors $\langle \Psi_\mathrm{I}(-\infty)|, |\Psi_\mathrm{I}(-\infty)\rangle$. For $t = 0$ the perturbation term H_1 is fully included and the system is in the perturbed state with the wave vectors $\langle \Psi_\mathrm{I}(t = 0)|, |\Psi_\mathrm{I}(t = 0)\rangle$. The adiabatic switching on of the perturbation term has to be very slow. After the evaluation has been performed we put $s \to +0$. This method of including perturbation is referred to as the adiabatic hypothesis.

The adiabatic switching on may be described in the form $H_1 \exp(-s|t|)$ too. For $t < 0$ the perturbation is gradually switched on and for $t > 0$ the perturbation is switched off.

The perturbed wave vectors may be determined using the operator of dynamic evolution $U(t, t')$ which transforms the wave vectors (the wave functions) given at the time t' into the wave vectors at the time t. Therefore, the wave vector at the time t is given by

$$\Psi_\mathrm{I}(t) = U(t, t')\, \Psi_\mathrm{I}(t') \,. \tag{13.21}$$

From Eq. (13.21) and from Eq. (13.13) we get

$$\mathrm{i}\hbar\frac{\partial}{\partial t}\left[U(t,\, t')\,\Psi_{\mathrm{I}}(t')\right] = H_{\mathrm{I}}(t)U(t,\, t')\,\Psi_{\mathrm{I}}(t')\ . \tag{13.22}$$

From Eq. (13.22) one may derive the operator equation (the equation of motion of the operator $U(t, t')$)

$$\mathrm{i}\hbar\frac{\partial U(t,\, t')}{\partial t} = H_{\mathrm{I}}(t)U(t,\, t') \tag{13.23}$$

which may be written in the integral form

$$U(t,\, t') = 1 - \frac{\mathrm{i}}{\hbar}\int_{t'}^{t} H_{\mathrm{I}}(t_1)U(t_1,\, t')\mathrm{d}t_1\ . \tag{13.24}$$

Using iteration method of evaluation, i.e. replacing $U(t_1,\, t')$ in Eq. (13.24) by the relation

$$U(t_1,\, t') = 1 - \frac{\mathrm{i}}{\hbar}\int_{t'}^{t_1} H_{\mathrm{I}}(t_2)U(t_2,\, t')\mathrm{d}t_2 \tag{13.25}$$

we get

$$\begin{aligned} U(t,\, t') &= 1 - \frac{\mathrm{i}}{\hbar}\int_{t'}^{t} H_{\mathrm{I}}(t_1)\left[1 - \frac{\mathrm{i}}{\hbar}\int_{t'}^{t_1} H_{\mathrm{I}}(t_2)U(t_2,\, t')\mathrm{d}t_2\right]\mathrm{d}t_1 \\ &= 1 - \frac{\mathrm{i}}{\hbar}\int_{t'}^{t} H_{\mathrm{I}}(t_1)\mathrm{d}t_1 + \left(-\frac{\mathrm{i}}{\hbar}\right)^2\int_{t'}^{t} H_{\mathrm{I}}(t_1)\mathrm{d}t_1\int_{t'}^{t_1} H_{\mathrm{I}}(t_2)U(t_2,\, t')\mathrm{d}t_2\ . \end{aligned} \tag{13.26}$$

This procedure may be repeated to derive a general formula for the operator of dynamic evolution.

13.4 Contour Integrals

We have to evaluate the integrals of the type

$$\begin{aligned} I_1 &= \int_{-\infty}^{+\infty}\frac{\exp(\mathrm{i}zt)}{z - \mathrm{i}s}\mathrm{d}z \qquad &(\mathrm{a}) \\ I_2 &= -\int_{-\infty}^{+\infty}\frac{\exp(-\mathrm{i}zt)}{z + \mathrm{i}s}\mathrm{d}z \qquad &(\mathrm{b}) \end{aligned} \tag{13.27}$$

where the integrated functions have singular points (simple poles as will be shown in the following) at $z = \mathrm{i}s$ and $z = -\mathrm{i}s$, respectively. We assume that the quantity s in Eq. (13.27) is a finite constant ($s \ll 1$, $\exp(st) \approx 1$ for a finite t). The integrals may be evaluated by application of the theory of residues. As the functions

$$\frac{\exp(\mathrm{i}zt)}{z - \mathrm{i}s} \quad \text{and} \quad \frac{\exp(-\mathrm{i}zt)}{z + \mathrm{i}s} \tag{13.28}$$

in the relations in (13.27) are continuous over the integration contours, analytic in the assumed region (except for the singular points) and the functions

$$(z - \mathrm{i}s)^{-1} \text{ and } (z + \mathrm{i}s)^{-1}$$

satisfy the condition of the uniform convergence the integrals may be evaluated by integration over a closed contours given in Fig. 13.1a and 13.1b, respectively. Therefore, we add to the integration contours over the real axis semicircles in the upper half-plane for evaluation of I_1 – Eq. (13.27a), and in the lower half-plane for evaluation of I_2 – Eq. (13.27b), i.e. above and below the real axis with the centres at the origin. The semicircles of the closed contours have the radii $R \to \infty$.

The first integral in Eq. (13.27) can be evaluated using the relation

$$I_1' = \underbrace{\int_{-\infty}^{+\infty} \frac{\exp(\mathrm{i}zt)}{z - \mathrm{i}s} \mathrm{d}z}_{I_1} + \underbrace{\int_R \frac{\exp(\mathrm{i}zt)}{z - \mathrm{i}s} \mathrm{d}z}_{I_R} . \tag{13.29}$$

First of all we will prove that the second integral (I_R) on the right hand

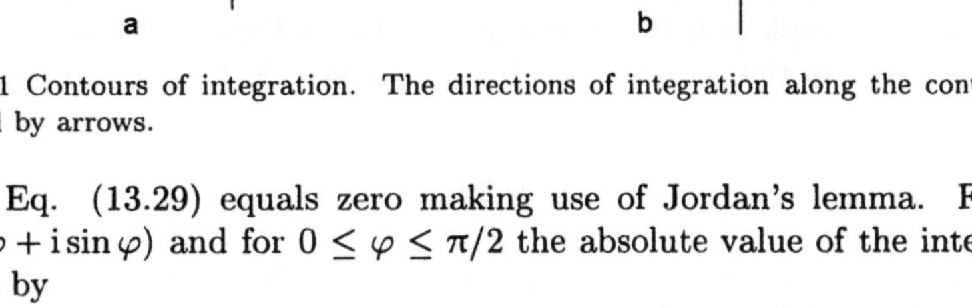

Fig. 13.1 Contours of integration. The directions of integration along the contours are indicated by arrows.

side of Eq. (13.29) equals zero making use of Jordan's lemma. For $z = R(\cos\varphi + \mathrm{i}\sin\varphi)$ and for $0 \le \varphi \le \pi/2$ the absolute value of the integral I_R is given by

$$|I_R| = \int_0^{\frac{\pi}{2}} \frac{1}{R - \Delta R} \exp(-tR\sin\varphi) R \mathrm{d}\varphi$$

$$\leq \frac{R}{R-\Delta R}\int_0^{\frac{\pi}{2}} \exp\left(-\frac{2tR}{\pi}\varphi\right)\mathrm{d}\varphi$$
$$= \frac{\pi}{2t(R-\Delta R)}\left[1-\exp\left(-tR\right)\right] \qquad (13.30)$$

where $R - \Delta R \leq |z - \mathrm{i}s|$ (ΔR is a small finite quantity) and the inequality

$$1 \geq \frac{\sin\varphi}{\varphi} \geq \frac{2}{\pi}$$

was used. For $R \to \infty$ the integral given by Eq. (13.30) vanishes. Likewise, the integral along the remaining part of the semicircle (for $\pi/2 \leq \varphi \leq \pi$) vanishes. The integral $I_R = 0$ and $I_1' = I_1$.

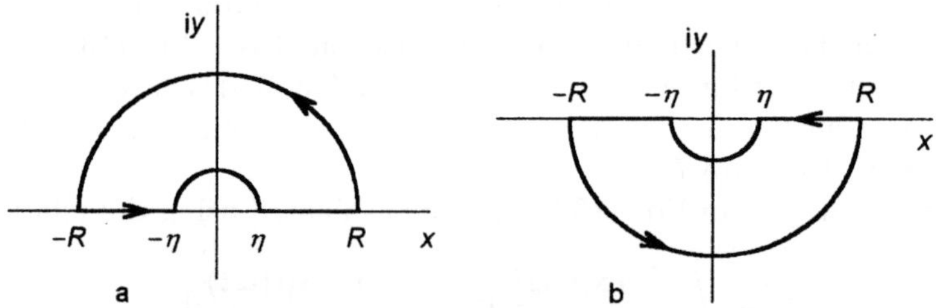

Fig. 13.2 Contours of integration. The directions of integration along the contours are indicated by arrows.

Now we determine the residue of the first function given by Eq. (13.28). In the neighborhood of the point $z = \mathrm{i}s$ this function may be expanded into the Laurent series

$$\frac{1}{z-\mathrm{i}s} + \mathrm{i}t + \frac{(\mathrm{i}t)^2}{2!}(z-\mathrm{i}s) + \frac{(\mathrm{i}t)^3}{3!}(z-\mathrm{i}s)^2 + \ldots \,. \qquad (13.31)$$

As the highest negative power of the series given by Eq. (13.31) is -1 the first function in Eq. (13.28) has the pole of the first order. The residue of the function at this simple pole is determined by the coefficient of the term whose power is -1, i.e. in this case the residue at the pole is 1. The integral in Eq. (13.27)a is

$$I_1 = 2\pi\mathrm{i}\,.$$

Likewise, using a similar procedure we may evaluate the second integral I_2 in Eq. (13.27). We find that

$$I_2 = 2\pi\mathrm{i}\,.$$

If $s \to 0$ the principal values of the integrals I_1 and I_2 in Eq. (13.27) may be evaluated. We will evaluate the limits

$$I_1 = \lim_{s \to +0} \int_{-\infty}^{+\infty} \frac{\exp(\mathrm{i}zt)}{z - \mathrm{i}s} \mathrm{d}z \,, \quad I_2 = - \lim_{s \to +0} \int_{-\infty}^{+\infty} \frac{\exp(-\mathrm{i}zt)}{z + \mathrm{i}s} \mathrm{d}z \,. \qquad (13.32)$$

To evaluate the integrals in Eq. (13.32) a method of going round the pole at $z = 0$ will be used. We alter the integration contour over the real axis. We will integrate over the negative real axis from $-R$ to $-\eta$, over a small semicircle in the upper (or lower) half-plane from $-\eta$ to $+\eta$, and over the positive real axis from $+\eta$ to R. The countour of integration is shown in Fig. 13.2. The integral I_1 is given by the relation

$$\lim_{s \to +0} \int_{-\infty}^{+\infty} \frac{\exp(\mathrm{i}zt)}{z - \mathrm{i}s} \mathrm{d}z = \lim_{s \to +0} \left[\int_{-\infty}^{-\eta} \frac{\exp(\mathrm{i}zt)}{z - \mathrm{i}s} \mathrm{d}z + \int_{+\eta}^{+\infty} \frac{\exp(\mathrm{i}zt)}{z - \mathrm{i}s} \mathrm{d}z + \int_{-\eta}^{+\eta} \frac{\exp(\mathrm{i}zt)}{z - \mathrm{i}s} \mathrm{d}z \right] . \qquad (13.33)$$

For $\eta \to 0$ the first two integrals on the right hand side of Eq. (13.33) determine the principal value of the integral I_1 which is written in the form

$$\mathcal{P} \int \frac{\exp(\mathrm{i}zt)}{z} \mathrm{d}z = \lim_{\substack{s \to +0 \\ \eta \to +0}} \left[\int_{-\infty}^{-\eta} \frac{\exp(\mathrm{i}zt)}{z - \mathrm{i}s} \mathrm{d}z + \int_{+\eta}^{+\infty} \frac{\exp(\mathrm{i}zt)}{z - \mathrm{i}s} \mathrm{d}z \right] . \qquad (13.34)$$

We will determine the third integral on the right hand side of Eq. (13.33). As the integrand function has a simple pole we may write it in the form

$$\frac{\exp(\mathrm{i}zt)}{z - \mathrm{i}s} = \frac{\exp(\mathrm{i}zt) - 1}{z - \mathrm{i}s} + \frac{1}{z - \mathrm{i}s} = G(z) + \frac{1}{z - \mathrm{i}s} \,. \qquad (13.35)$$

The integral over the small semicircle where $z - is = \eta \exp(\mathrm{i}\varphi)$ is given by

$$\lim_{\substack{s \to +0 \\ \eta \to +0}} \int_{\pi}^{0} \left[G(z) + \frac{1}{z - \mathrm{i}s} \right] \eta \mathrm{i} \exp(\mathrm{i}\varphi) \mathrm{d}\varphi$$

$$= \lim_{\substack{s \to +0 \\ \eta \to +0}} \int_{\pi}^{0} [\eta G(z) + \exp(-\mathrm{i}\varphi)] \, \mathrm{i} \exp(\mathrm{i}\varphi) \mathrm{d}\varphi = -\pi \mathrm{i} \,. \qquad (13.36)$$

As $I_R = 0$ for $R \to \infty$ (which follows from Eq. (13.30)) the principal value of the integral I_1 in Eq. (13.27) may be determined from Eqs (13.33), (13.34), and (13.36). We get

$$\mathcal{P} \int \frac{\exp(\mathrm{i}zt)}{z - \mathrm{i}s} \mathrm{d}z = \pi \mathrm{i} \,. \qquad (13.37)$$

Likewise, the principal value of the integral I_2 may be found. In this case we may use a small semicircle in the lower half-plane from $+\eta$ to $-\eta$ to go round the pole at $z = 0$ (Fig. 13.2). The principal value of the integral I_2 in Eq. (13.27) is given by the relation

$$-\mathcal{P}\int \frac{\exp(-\mathrm{i}zt)}{z+\mathrm{i}s}\mathrm{d}z = \pi\mathrm{i}\,. \tag{13.38}$$

13.5 Stochastic Differential Equation and Transients in Linear Systems

Solution of the stochastic differential equation

$$\ddot{y}(t) + \sigma\dot{y}(t) + \omega_0^2 y(t) = x(t) \tag{13.39}$$

where $x(t)$ is the stochastic stationary process and ω_0, $\sigma = (\beta - \gamma)$ are constants (β and γ are the positive damping and the negative damping, respectively). Solution of the stochastic differential equation (13.39) can be considered as the output $y(t)$ of a linear system with the input $x(t)$ and the system function [136]

$$F(\mathrm{i}\,\omega) = \frac{1}{(\mathrm{i}\,\omega)^2 + \mathrm{i}\,\omega\sigma + \omega_0^2}\,. \tag{13.40}$$

We will assume that $\sigma > 0$. If the process $x(t)$ is stationary then the stochastic differential equation is valid for any time and the process $y(t)$ is stationary too. We assume that $x(t)$ has a power spectrum $S_{xx}(\omega)$. The power spectrum of $y(t)$ is given by the relation

$$S_{yy}(\omega) = S_{xx}(\omega)\,|F(\mathrm{i}\,\omega)|^2 = \frac{S_{xx}(\omega)}{(\omega^2-\omega_0^2)^2+\sigma^2\omega^2}\,. \tag{13.41}$$

The autocorrelation $R_{yy}(\tau)$ and the power spectrum $S_{yy}(\omega)$ are mutually related by

$$R_{yy}(\tau) = \frac{1}{2\pi}\int_{-\infty}^{+\infty} S_{yy}(\omega)\exp(\mathrm{i}\,\omega\tau)\mathrm{d}\omega \tag{13.42}$$

$$S_{yy}(\omega) = \int_{-\infty}^{+\infty} R_{yy}(\tau)\exp(-\mathrm{i}\,\omega\tau)\mathrm{d}\tau\,. \tag{13.43}$$

From Eqs (13.41) and (13.42) one gets

$$R_{yy}(\tau) = \frac{1}{2\pi}\int_{-\infty}^{+\infty} \frac{S_{xx}(\omega)}{(\omega^2-\omega_0^2)^2+\sigma^2\omega^2}\exp(\mathrm{i}\,\omega\tau)\mathrm{d}\omega\,. \tag{13.44}$$

We assume that the power spectrum $S_{xx}(\omega)$ is flat and does not depend on ω, i.e. $S_{xx}(\omega) = \alpha\kappa$ where $\alpha\kappa$ is a constant. After substitution of p/i for ω into Eq. (13.44) we obtain

$$\begin{aligned} R_{yy}(\tau) &= \frac{\alpha\kappa}{2\pi\mathrm{i}} \int_{-\mathrm{i}\infty}^{+\mathrm{i}\infty} \frac{\exp(p\tau)}{(p^2+\omega_0^2)^2 - \sigma^2 p^2} \mathrm{d}p \\ &= \frac{\alpha\kappa}{2\pi\mathrm{i}} \int_{-\mathrm{i}\infty}^{+\mathrm{i}\infty} \frac{1}{2\sigma\omega_0^2} \left[\frac{p+\sigma}{p^2+\omega_0^2+\sigma p} + \frac{-p+\sigma}{p^2+\omega_0^2-\sigma p} \right] \exp(p\tau)\mathrm{d}p \,. \end{aligned} \tag{13.45}$$

The bilateral Laplace transform is used in [136] to determine $R_{yy}(\tau)$ in Eq. (13.45). The first term on the right hand side of Eq. (13.45) has no singular points in the plane $\Re(p) > 0$. The inverse transform of this term equals zero for $\tau < 0$. Changing p to $-p$ we get the second term on the right hand side of Eq. (13.45) which has no singular points in the plane $\Re(p) < 0$. The inverse transform of this term equals zero for $\tau > 0$ (and yields a non zero result for $\tau < 0$). Therefore, it is sufficient to find the solution for $\tau > 0$. The first term on the right hand side of Eq. (13.45) can be written in the form

$$\frac{p+\sigma}{p^2+\omega_0^2+\sigma p} = \frac{p+\frac{\sigma}{2}}{\left(p+\frac{\sigma}{2}\right)^2+\omega_0^2-\frac{\sigma^2}{4}} + \frac{\frac{\sigma}{2}}{\sqrt{\omega_0^2-\frac{\sigma^2}{4}}} \frac{\sqrt{\omega_0^2-\frac{\sigma^2}{4}}}{\left(p+\frac{\sigma}{2}\right)^2+\omega_0^2-\frac{\sigma^2}{4}} \,. \tag{13.46}$$

The inverse Laplace transform of the terms on the right hand side of Eq. (13.46) depends on the value $\omega^2 = \omega_0^2 - \sigma^2/4$ which may be positive ($\omega_0^2 > \sigma^2/4$), negative ($\omega_0^2 < \sigma^2/4$), or may equal zero ($\omega_0^2 = \sigma^2/4$). The inverse transforms for all three cases which are given by Eqs (6.6)–(6.8) are elementary.

In a system where the damping is balanced with the negative resistance σ is zero and the system function is given by the relation

$$F(\mathrm{i}\omega) = \frac{1}{(\mathrm{i}\omega)^2+\omega_0^2} \,. \tag{13.47}$$

In this case the system is not damped and we may expect that the process is not stationary (in the wide sense). We will use a method of transients [136] to assess the behaviour of the system. The solution will be limited to a certain time interval. The autocorrelation of a stochastic stationary process $x(t)$ (where $x(t)$ is the white noise) is given by

$$\begin{aligned} R_{xx}(\tau) &= R_{xx}(t_1,t_2) = \alpha\kappa\delta(t_1-t_2) \qquad &&\text{for} \quad 0 \le t_1,\ 0 \le t_2 \\ R_{xx}(\tau) &= R_{xx}(t_1,t_2) = 0 &&\text{otherwise} \end{aligned} \tag{13.48}$$

where $\tau = t_1 - t_2$. The impulse response is given by

$$f(t) = \frac{1}{\omega_0} \sin(\omega_0 t) U(t) \tag{13.49}$$

where $U(t)$ is the unit step function ($U(t) = 1$ for $0 \le t$ and $U(t) = 0$ for $t < 0$). Using Eqs (13.48) and (13.49) the cross correlation $R_{yx}(t_1, t_2)$ between the input and the output is evaluated from

$$\begin{aligned} R_{yx}(\tau) = R_{yx}(t_1, t_2) &= \int_{-\infty}^{+\infty} R_{xx}(t_1 - t', t_2) f(t') \mathrm{d}t' \\ &= \frac{\alpha\kappa}{\omega_0} \sin[\omega(t_1 - t_2)] U(t_1 - t_2) \qquad \text{for} \quad 0 \le t_1,\ 0 \le t_2\,. \end{aligned} \tag{13.50}$$

The autocorrelation of the output is given by

$$\begin{aligned} R_{yy}(\tau) = R_{yy}(t_1, t_2) &= \int_{-\infty}^{+\infty} R_{yx}(t_1, t_2 - t') f(t') \mathrm{d}t' \\ &= \frac{\alpha\kappa}{2\omega_0^2} \left\{ \sin[\omega_0(t_1 - t_2)] \frac{\sin^2(\omega_0 t_2)}{\omega_0} \right. \\ &\quad \left. + \cos[\omega_0(t_1 - t_2)] \left[t_2 - \frac{\cos(\omega_0 t_2) \sin(\omega_0 t_2)}{\omega_0} \right] \right\} \end{aligned} \tag{13.51}$$

where $0 \le t_1$, $0 \le t_2$, and $t_1 \ge t_2$.

A similar method may be used for $\sigma < 0$. We will assume that $|\sigma|$ is small and that $|\sigma| < \omega_0$. The system function is given by

$$F(\mathrm{i}\omega) = \frac{1}{(\mathrm{i}\omega)^2 - \mathrm{i}\omega|\sigma| + \omega_0^2} \tag{13.52}$$

and the impulse response

$$f(t) = \frac{1}{\omega_1} \exp\left(\frac{|\sigma|}{2} t\right) \sin(\omega_1 t) U(t) \tag{13.53}$$

where $\omega_1 = \sqrt{\omega_0^2 - (\sigma/2)^2}$.

Using $f(t)$ from Eq. (13.53) instead of $f(t)$ given by Eq. (13.49) we obtain from Eq. (13.50) $R_{yx}(t_1, t_2)$ and further from Eq. (13.51) $R_{yy}(t_1, t_2)$ in the form

$$\begin{aligned} R_{yy}(\tau) &= R_{yy}(t_1, t_2) \\ &= \frac{\alpha\kappa}{\omega_1^2} \exp\left[\frac{|\sigma|}{2}(t_1 - t_2)\right] \{\sin[\omega_1(t_1 - t_2)] \rho_\mathrm{s}(t_2) \\ &\quad + \cos[\omega_1(t_1 - t_2)] \rho_\mathrm{c}(t_2)\} \end{aligned} \tag{13.54}$$

where

$$\begin{aligned}\rho_s(t_2) &= \frac{1}{4}\frac{\omega_1}{\omega_0^2}\left\{\exp(|\sigma|t_2)\left[\frac{|\sigma|}{2\omega_1}\sin(2\omega_1 t_2) - \cos 2\omega_1 t_2\right] + 1\right\}\\ \rho_c(t_2) &= \frac{1}{2|\sigma|}\left[\exp(|\sigma|t_2) - 1\right]\\ &\quad -\frac{1}{8}\frac{|\sigma|}{\omega_0^2}\left\{\exp(|\sigma|t_2)\left[\frac{2\omega_1}{|\sigma|}\sin(2\omega_1 t_2) + \cos(2\omega_1 t_2)\right] - 1\right\}\end{aligned}$$

for $0 \le t_1,\ 0 \le t_2,$ and $t_1 \ge t_2$. (13.55)

The derived relations are used to analyze the coherence in the vibration systems with energy condensation.

13.6 Some Relations from Vector Analysis

In Section 8.2 we use formulas from vector analysis in the spherical coordinates. The spherical coordinates R, Θ, Φ are related to the rectangular coordinates x, y, z by

$$x = R\sin\Theta\cos\Phi, \quad y = R\sin\Theta\sin\Phi, \quad z = R\cos\Theta .$$

We introduce necessary relations of the vector analysis expressed in spherical coordinates.

The symbolic vector "nabla" is a vector operator given by

$$\nabla = \frac{\partial}{\partial x}\boldsymbol{i} + \frac{\partial}{\partial y}\boldsymbol{j} + \frac{\partial}{\partial z}\boldsymbol{k} = \frac{\partial}{\partial R}\boldsymbol{r} + \frac{1}{R}\frac{\partial}{\partial\Theta}\boldsymbol{\vartheta} + \frac{1}{R\sin\Theta}\frac{\partial}{\partial\Phi}\boldsymbol{\varphi} \qquad (13.56)$$

where $\boldsymbol{i}$, $\boldsymbol{j}$, $\boldsymbol{k}$ and $\boldsymbol{r}$, $\boldsymbol{\vartheta}$, $\boldsymbol{\varphi}$ are the unit vectors in the rectangular coordinates and in the spherical coordinates, respectively.

The gradient defined in a scalar field and determining the greatest change of the scalar quatity Ψ and its direction in space is obtained from the relation

$$\nabla\Psi = \frac{\partial\Psi}{\partial R}\boldsymbol{r} + \frac{1}{R}\frac{\partial\Psi}{\partial\Theta}\boldsymbol{\vartheta} + \frac{1}{R\sin\Theta}\frac{\partial\Psi}{\partial\Phi}\boldsymbol{\varphi} . \qquad (13.57)$$

The divergence as a measure of the change of a vector quantity $\boldsymbol{M}$ in an infinitesimal volume (and referred to the unit volume) related by Gauss's Theorem to the flow over a surface enclosing the volume is given by the relation

$$\nabla\cdot\boldsymbol{M} = \frac{1}{R^2}\frac{\partial}{\partial R}\left(R^2 M_R\right) + \frac{1}{R\sin\Theta}\frac{\partial}{\partial\Theta}(\sin\Theta M_\Theta) + \frac{1}{R\sin\Theta}\frac{\partial M_\Phi}{\partial\Phi} . \qquad (13.58)$$

The curl of a vector as a measure of the flow across an infinitesimal area (and referred to the unit area) related by Stokes's Theorem to the line integral of the vector taken over a closed curve which is a boundary of the area is expressed by

$$\nabla \times \boldsymbol{M} = \begin{vmatrix} \frac{\boldsymbol{r}}{R^2 \sin \Theta} , & \frac{\boldsymbol{\vartheta}}{R \sin \Theta} , & \frac{\boldsymbol{\varphi}}{R} \\ \frac{\partial}{\partial R} , & \frac{\partial}{\partial \Theta} , & \frac{\partial}{\partial \Phi} \\ M_R , & RM_\Theta , & R \sin \Theta M_\Phi \end{vmatrix}$$

$$= \frac{1}{R \sin \Theta} \left[\frac{\partial}{\partial \Theta} (\sin \Theta M_\Phi) - \frac{\partial M_\Theta}{\partial \Phi} \right] \boldsymbol{r}$$
$$+ \frac{1}{R} \left[\frac{1}{\sin \Theta} \frac{\partial M_R}{\partial \Phi} - \frac{\partial}{\partial R} (RM_\Phi) \right] \boldsymbol{\vartheta}$$
$$+ \frac{1}{R} \left[\frac{\partial}{\partial R} (RM_\Theta) - \frac{\partial M_R}{\partial \Theta} \right] \boldsymbol{\varphi} . \quad (13.59)$$

The vector product of the type $\nabla \times \nabla \times \boldsymbol{M}$ may be easily derived from Eq. (13.59). We replace the components M_R, M_Θ, M_Φ in the last row of the determinant, respectively, by the components of $\nabla \times \boldsymbol{M}$ in the directions $\boldsymbol{r}$, $\boldsymbol{\vartheta}$, $\boldsymbol{\varphi}$ from the right hand side of Eq. (13.59).

The relation containing the Laplacian operator ∇^2 acting on a scalar function Ψ is given by

$$\nabla^2 \Psi = \nabla \cdot \nabla \Psi = \frac{1}{R^2} \frac{\partial}{\partial R} \left(R^2 \frac{\partial \Psi}{\partial R} \right)$$
$$+ \frac{1}{R^2 \sin \Theta} \frac{\partial}{\partial \Theta} \left(\sin \Theta \frac{\partial \Psi}{\partial \Theta} \right) + \frac{1}{R^2 \sin^2 \Theta} \frac{\partial^2 \Psi}{\partial \Phi^2} . \quad (13.60)$$

By the operator ∇^2 acting on a vector one must understand [146]

$$\nabla^2 \boldsymbol{M} = \nabla \nabla \cdot \boldsymbol{M} - \nabla \times \nabla \times \boldsymbol{M} . \quad (13.61)$$

and the vector wave equation is given by

$$\nabla^2 \boldsymbol{M} + k^2 \boldsymbol{M} = \nabla \nabla \cdot \boldsymbol{M} - \nabla \times \nabla \times \boldsymbol{M} + k^2 \boldsymbol{M} \quad (13.62)$$

where k is the propagation constant.

13.7 Electromagnetic Field of the Electric Dipole

A solution of the inhomogeneous scalar wave equation enable calculations of the electromagnetic field from a given distribution of charges and currents.

Using a multipole expansion the relations for electric and magnetic dipoles and multipoles are derived. The oscillating electric dipole generates the electromagnetic field which is described by the well known relations [146]

$$
\begin{aligned}
E_R &= \frac{1}{2\pi\varepsilon}\left(\frac{1}{R^3} - \frac{\mathrm{i}k}{R^2}\right)\cos\Theta|p|\exp(\mathrm{i}\omega t^*) \\
E_\Theta &= \frac{1}{4\pi\varepsilon}\left(\frac{1}{R^3} - \frac{\mathrm{i}k}{R^2} - \frac{k^2}{R}\right)\sin\Theta|p|\exp(\mathrm{i}\omega t^*) \\
H_\Phi &= -\frac{\mathrm{i}\omega}{4\pi}\left(\frac{1}{R^2} - \frac{\mathrm{i}k}{R}\right)\sin\Theta|p|\exp(\mathrm{i}\omega t^*)
\end{aligned}
\tag{13.63}
$$

where $t^* = t - kR$ is the retarded time variable, E_R, E_Θ, and H_Φ are the intensities of the electric and of the magnetic field in the spherical coordinates R, Θ, and Φ, respectively. The permittivity $\varepsilon = \varepsilon'\varepsilon_0$ where ε' is the relative permittivity and ε_0 is the permittivity of the free space. k is given by the relation $k^2 = \varepsilon\mu\omega^2 + \mathrm{i}\mu\sigma\omega$ where μ is the permeability and σ is the conductivity.

We note that the electric field vectors lie in a meridian plane through the axis of the dipole and the magnetic vector is perpendicular to this plane and forms a tangent to coaxial circles about the dipole axis. In the immediate neighbourhood of the source the "static" $(1/R^3)$ and "induction" $(1/R^2)$ fields are dominant while at distances much greater than the wavelength the "radiation" field $(1/R)$ is important. In the vicinity of the source there is a longitudinal component of the intensity of the electric field in the direction of propagation. At great distances from the dipole the field is transverse to the direction of propagation. Energy flows from the dipole which loses energy at a constant rate. Using the complex Poynting vector

$$
\boldsymbol{S}^* = \frac{1}{2}(\boldsymbol{E} \times \boldsymbol{H}^*) \tag{13.64}
$$

the mean value of the energy per second crossing a unit area in the outwards perpendicular direction is given by its real part

$$
P = \Re(\boldsymbol{S}^*) = \frac{1}{2}\Re(E_\Theta H_\Phi^*) = \frac{\omega k^3}{32\pi^2\varepsilon}\frac{\sin^2\Theta}{R^2}|p|^2 \tag{13.65}
$$

where the right hand side was derived using Eqs (13.63). Integrating the right hand side term of Eq. (13.65) over the surface of a sphere of the radius R one gets

$$
\int_{\Phi=0}^{2\pi}\int_{\Theta=0}^{\pi}\Re(\boldsymbol{S}^*)\,R^2\sin\Theta\,\mathrm{d}\Theta\,\mathrm{d}\Phi = \frac{\omega^4}{12\pi}\mu\sqrt{\mu\varepsilon}|p|^2 \tag{13.66}
$$

for $k^2 = \mu\varepsilon\omega^2$.

As the great majority of protein macromolecules and polymer structures are polar the electric dipole or multipole fields are important for assessment of interactions in living systems.

13.8 Probability and Information

A measure of information is based on probabilty. Probability of an event a is a number $P(a)$ assigned to this event. This number is positive ($P(a) \geq 0$), its greatest value $P(a) = 1$ for a certain event, and mutually exclusive events a and b have $P(a+b) = P(a) + P(b)$. The values of these numbers P may be determined experimentally from the relative frequency of occurrence of the events as a limit of the ratio of the positive outcomes n_a (i.e. how many times the event occurred) to the total number n of the experiments, therefore,

$$P(a) = \lim_{n \to \infty} n_a / n$$

or may be found a priori by evaluating the ratio of the favourable alternatives to the total number of the possible alternatives. Discussion about the probability definitions can be found e.g. in [136].

One may also introduce an important concept of the conditional probability $P(b/a)$ which is the probability that the event b will occur if the event a is known to have occurred. The formal definition of the conditional probability is given by the relation

$$P(b/a) = \frac{P(a,b)}{P(a)} \tag{13.67}$$

where $P(a,b)$ is the probability that both events occur. Likewise,

$$P(a/b) = \frac{P(a,b)}{P(b)} \,. \tag{13.68}$$

In Eqs (13.67) and (13.68) $P(a,b) = P(b,a)$ but $P(a/b) \neq P(b/a)$. For independent events $P(a,b) = P(a)P(b)$.

Now we can describe the measure of information. If something happens contrary to all expectation the information content is high. An event with high probability carries negligible information and a certain event none. Information of the event a depends on the reciprocal value of the probability $P(a)$ that this event occurs. Information is defined as

$$I = \lg \frac{1}{P(a)} \,. \tag{13.69}$$

On the basis of the definition the unit of information was determined as the binary unit – bit.

We assume that there is an ensemble A of events a_m. The information entropy $\langle I \rangle$ is the expectation value (the average value) of information given on the ensemble of events

$$\langle I \rangle = \sum_m P(a_m) \lg \frac{1}{P(a_m)} . \tag{13.70}$$

We will derive a condition for the maximum value of the information entropy. The number of different events a_m in the ensemble A is N. We will compare the entropy $\langle I \rangle$ of a system in which the probabilities $P(a_m)$ of individual events are different with the entropy of a similar system where all the events are equiprobable, i.e. $P = 1/N$. Under the equiprobability condition the information entropy of the system is given by the relation

$$\langle I_{\mathrm{eq}} \rangle = \sum \frac{1}{N} \lg N = \lg N \tag{13.71}$$

as follows from Eq. (13.70). We will evaluate the difference [158]

$$\begin{aligned} \langle I \rangle - \langle I_{\mathrm{eq}} \rangle &= \sum_m P(a_m) \lg \frac{1}{P(a_m)} - \lg N \\ &= \sum_m P(a_m) \lg \frac{1}{P(a_m)} + \sum_m P(a_m) \lg \frac{1}{N} \\ &= \sum_m P(a_m) \lg \frac{1}{P(a_m)N} \leq \sum_m \left[\frac{1}{N} - P(a_m) \right] = 0 \end{aligned} \tag{13.72}$$

where the relations

$$\sum_m P(a_m) = 1 \quad \text{and} \quad \lg \frac{1}{P(a_m)N} \leq \frac{1}{P(a_m)N} - 1$$

were used. Equation (13.72) proves that information entropy reaches the maximum value when the probabilities of all events in the ensemble are equal. Under condition of the maximum entropy the amount of information provided is the information capacity.

Transfer of information through a channel may be analyzed using the concept of mutual information defined as

$$I(a_m, b_n) = \lg \frac{P(a_m, b_n)}{P(a_m)P(b_n)} \tag{13.73}$$

where a_m is the input event from the input ensemble A and b_n is the output event from the output ensemble B (Fig. 13.3).

Using relations (13.67) and (13.68) for the conditional probability Eq. (13.73) can be written in the form

$$I(a_m, b_n) = \lg \frac{P(a_m/b_n)}{P(a_m)} = \lg \frac{P(b_n/a_m)}{P(b_n)} . \qquad (13.74)$$

The input information entropy of the ensemble A at the input of the com-

Fig. 13.3 A communication channel.

munication channel and the output information entropy of the ensamble B (i.e. the average input and the average ouput selfinformation, respectively) are given by the relations

$$\langle I(A) \rangle = \sum_m P(a_m) \lg \frac{1}{P(a_m)} , \quad \langle I(B) \rangle = \sum_n P(b_n) \lg \frac{1}{P(b_n)} . \qquad (13.75)$$

The average mutual information is derived from Eq. (13.73). One obtains

$$I(A; B) = \sum_m \sum_n P(a_m, b_n) \lg \frac{P(a_m/b_n)}{P(a_m)} . \qquad (13.76)$$

Using the theorem of total probability [136]

$$\sum_n P(a_m, b_n) = \sum_n P(a_m/b_n) P(b_n) = P(a_m)$$

one obtains from Eqs (13.75) and (13.76) for the average mutual information

$$\begin{aligned} I(A; B) &= \sum_m \sum_n P(a_m, b_n) \lg P(a_m/b_n) - \sum_m \sum_n P(a_m, b_n) \lg P(a_m) \\ &= \langle I(A) \rangle - \langle I(A/B) \rangle . \end{aligned} \qquad (13.77)$$

Likewise, using the relation $P(a_m, b_n) = P(b_n, a_m) = P(b_n/a_m)P(a_m)$ which follows from Eq. (13.67) one obtains from Eq. (13.76)

$$\begin{aligned} I(B; A) &= \sum_m \sum_n P(b_n, a_m) \lg P(b_n/a_m) - \sum_m \sum_n P(b_n, a_m) \lg P(b_n) \\ &= \langle I(B) \rangle - \langle I(B/A) \rangle . \end{aligned} \qquad (13.78)$$

$\langle I(A)\rangle$ in Eq. (13.77) is considered to be the average amount of information at the input of the channel and $\langle I(A/B)\rangle$ is the average amount of information loss in the channel (e.g. due to noise). The conditional entropy $\langle I(A/B)\rangle$ is usually regarded as the equivocation of the channel. The information entropy $\langle I(B)\rangle$ (Eq. (13.78)) at the output of the channel is considered as the average amount of information received at the output of the channel and $\langle I(B/A)\rangle$ specifies the noise disturbance of the channel (the noise entropy).

13.9 Information and Physical Entropy

The Boltzmann relation, one of the most important equations in physics

$$\mathcal{S}_{\text{phys}} = k \ln P \tag{13.79}$$

relates the physical entropy $\mathcal{S}_{\text{phys}}$ to the thermodynamic probability P. The thermodynamic probability is a measure of the probability of occurrence of certain thermodynamic state. This probability is given by the number of micro-distributions (i.e. by the number of microstates) corresponding to a given macrostate. In the classical statistics of Boltzmann the thermodynamic probability is given by the number of ways in which the total number N of entities (e.g. molecules) is distributed among Z cells so that there are N_i entities in the i-th cell. The cells are equivalent and the equal a priori probability of each cell to accept an entity is not disturbed by the number N_i. Changing the positions of the entities within one cell does not yield a new microstate. The thermodynamic probability is given by

$$P = \frac{N!}{N_1!N_2!N_3!\ldots} = \frac{N!}{\prod^i N_i!} = \frac{N^N}{\prod^i N_i^{N_i}} \frac{\prod^i \exp(N_i)}{\exp(N)} = \frac{\prod^i N^{N_i}}{\prod^i N_i^{N_i}} \tag{13.80}$$

where we applied the Stirling formula

$$N! \approx \frac{N^N}{\exp(N)}\sqrt{2\pi N} \approx \frac{N^N}{\exp(N)}$$

for large N. The probability given by Eq. (13.80) enable us to determine the relation

$$-\ln P = \ln\frac{1}{P} = \ln\left(\frac{\prod^i N_i}{\prod^i N}\right)^{N_i} = N\sum_{i=0}^{i_{\max}} \frac{N_i}{N}\ln\frac{N_i}{N}\,. \tag{13.81}$$

We assume that the limit of the ratio N_i/N exists and that for large values of N and N_i this ratio is equal to the probability P_i, i.e.

$$P_i = \lim_{N\to\infty}\frac{N_i}{N}\,. \tag{13.82}$$

From Eqs (13.79)–(13.81) we get

$$\mathcal{S}_{\text{phys}} = -kN \sum_i P_i \ln P_i \ . \tag{13.83}$$

In Eq. (13.70) information entropy is defined as

$$\mathcal{S}_{\text{inf}} = \langle I \rangle = -\sum_i P_i \lg P_i \ .$$

Using this relation we get from Eq. (13.83) [37]

$$\mathcal{S}_{\text{phys}} = kN\mathcal{S}_{\text{inf}} \ln 2 \ . \tag{13.84}$$

We can deduce from Eq. (13.84) that the physical entropy and the information entropy describe the same processes in nature.

13.10 Channel Capacity Integral

To determine the capacity of the communication channel in Section 9.4 we have to evaluate the integral (Eq. (9.20))

$$\begin{aligned} C_i &= \frac{1}{k\ln 2}\int_{T_\text{n}}^{T_{\text{s}i}} \frac{\text{d}\langle E_{\text{n}i}(T)\rangle}{T} = \frac{1}{k\ln 2}\int_{T_\text{n}}^{T_{\text{s}i}} \frac{\text{d}\langle E_{\text{n}i}(T)\rangle}{\text{d}T}\frac{\text{d}T}{T} \\ &= \frac{1}{k\ln 2}\left[\frac{\langle E_{\text{o}i}(T_{\text{s}i})\rangle}{T_{\text{s}i}} - \frac{\langle E_{\text{n}i}(T_\text{n})\rangle}{T_\text{n}}\right] + \frac{1}{k\ln 2}\int_{T_\text{n}}^{T_{\text{s}i}} \frac{\langle E_{\text{n}i}(T)\rangle}{T^2}\text{d}T \end{aligned} \tag{13.85}$$

where $\langle E_{\text{n}i}(T_\text{n})\rangle = \langle E_{\text{n}i}\rangle$ and $\langle E_{\text{o}i}(T_\text{n})\rangle = \langle E_{\text{o}i}\rangle$ are given by the relations

$$\langle E_{\text{n}i}\rangle = \frac{h\nu_i}{\exp\left(\dfrac{h\nu_i}{kT_\text{n}}\right) - 1} \quad \text{and} \quad \langle E_{\text{o}i}\rangle = \frac{h\nu_i}{\exp\left(\dfrac{h\nu_i}{kT_{\text{s}i}}\right) - 1} = \langle E_{Di}\rangle + \langle E_{\text{n}i}\rangle \tag{13.86}$$

where T_n and $T_{\text{s}i}$ are the temperatures corresponding to the noise and to the signal, respectively. From the second relation in (13.86) one can determine the temperature $T_{\text{s}i}$ in the form

$$T_{\text{s}i} = \frac{1}{\dfrac{k}{h\nu_i}\ln\left(1 + \dfrac{h\nu_i}{\langle E_{Di}\rangle + \langle E_{\text{n}i}\rangle}\right)} \ . \tag{13.87}$$

We replace $\langle E_{ni}(T)\rangle$ in the integral on the right hand side of Eq. (13.85) by the first relation in Eq. (13.86) and then make use of the substitution

$$x = \frac{h\nu_i}{kT} \quad \text{and then} \quad y = \exp(x) \, .$$

One gets

$$\begin{aligned} \frac{1}{k}\int_{T_n}^{T_{si}} \frac{\langle E_{ni}\rangle}{T^2}\mathrm{d}T &= -\int_{T_n}^{T_{si}} \left(-\frac{h\nu_i}{kT^2}\right) \frac{1}{\exp\left(\frac{h\nu_i}{kT}\right) - 1}\mathrm{d}T \\ &= -\int_{x_1}^{x_2} \frac{\mathrm{d}x}{\exp(x) - 1} = -\int_{y_1}^{y_2} \frac{\mathrm{d}y}{y(y-1)} \\ &= \int_{y_1}^{y_2} \frac{\mathrm{d}y}{y} - \int_{y_1}^{y_2} \frac{\mathrm{d}y}{y-1} = \ln\left[\frac{y_2(y_1-1)}{y_1(y_2-1)}\right] \, . \end{aligned} \tag{13.88}$$

Equations (13.85) and (13.88) enable us to determine the capacity of the channel. Using Eqs (13.86) we get

$$C_i = \frac{1}{k\ln 2}\left[\frac{\langle E_{oi}\rangle}{T_{si}} - \frac{\langle E_{ni}\rangle}{T_n}\right] + \frac{1}{\ln 2}\ln\left[\frac{\left(1+\frac{h\nu_i}{\langle E_{oi}\rangle}\right)\frac{h\nu_i}{\langle E_{ni}\rangle}}{\left(1+\frac{h\nu_i}{\langle E_{ni}\rangle}\right)\frac{h\nu_i}{\langle E_{oi}\rangle}}\right] \, . \tag{13.89}$$

After substitution $\langle E_{oi}\rangle$, $\langle E_{ni}\rangle$, T_{si} from the relations (13.86) and (13.87) into Eq. (13.89) one gets the final relation given by Eq. (9.21) [158].

13.11 Energy Flow from the Electric Dipole

We assume the electric dipole field. The mean outward flow of energy is given by the real part of the complex Poynting vector given by Eq. (13.64) [146]. In the equatorial region we get the relation

$$\boldsymbol{S}_i = \frac{1}{2}\Re(\boldsymbol{E}_\Theta \times \boldsymbol{H}^*_\Phi)_i = \left(k_i^3\frac{\nu_i p_i^2 \sin^2\Theta}{16\pi\varepsilon_i R^2}\right)\boldsymbol{r} = G_i\nu_i^4\,\frac{\sin^2\Theta}{R^2}\boldsymbol{r} \tag{13.90}$$

and

$$G_i = \frac{\pi^2\mu p_i^2}{2}\sqrt{\mu\varepsilon_i}$$

where $k_i = 2\pi\nu_i\sqrt{\mu\varepsilon_i}$ is the propagation constant at the frequency ν_i (for a medium whose conductivity is zero), p_i is the dipole moment (scalar), ε_i is the permittivity of the medium, R, Θ, and Φ are the spherical coordinates, and $\boldsymbol{r}$ is the unit vector. (The contribution of the term $(\boldsymbol{E}_R \times \boldsymbol{H}^*_\Phi)$ is neglected.)

In the vicinity of the source the generated field has the "static" and the "inductive" components, which vary as $1/R^3$ and as $1/R^2$, respectively. At small values of R these near zone components are greater than the radiation component (whose Poynting vector is given by Eq. (13.90)), which varies as $1/R$. The maximum values of the energy densities of the near zone components in the direction of the dipole axis and in the direction perpendicular to it in the equatorial plane are given, respectively, by the relations

$$W_{iR} = \frac{1}{2}\varepsilon_i |E_{iR}|^2 = \frac{\cos^2\Theta}{R^2} g_i \tag{13.91}$$

$$W_{i\Theta} = \frac{1}{2}\varepsilon_i |E_{i\Theta}|^2 = \frac{\sin^2\Theta}{4R^2} g_i \,. \tag{13.92}$$

The coefficient g_i in Eqs (13.91) and (13.92) is given by the relation

$$g_i = p_i^2 \left(\frac{1}{8\pi^2 \varepsilon_i} \frac{1}{R^4} + \frac{\nu_i^2 \mu_i}{2} \frac{1}{R^2} \right)$$

where the "radiation" term $1/R$ of the intensity $E_{i\Theta}$ given by Eq. (13.63) was neglected.

The electromagnetic energy is distributed in the medium around the oscillating units. The radiation component flows outwards from it. The near zone components flow outwards from the source and then return back to it within one period of oscillation. A part of the energy falls on another oscillating unit of the Fröhlich system. The amount of energy absorbed by the oscillating unit depends on its effective cross-section and on the density of energy flow. We assume here that the effective cross-section is determined by the dimensions of the receiving oscillating unit (e.g. of protein molecules). We assume the cross-section to be a rectangle D whose sides are x and z. Using Eq. (13.90) the energy flow of the radiation component across the cross-section rectangle may be determined by the relation

$$\begin{aligned} \langle W_{iP} \rangle &= \int_D S_i \mathrm{d}D = 4G_i \nu_i^4 \int_0^{\Phi_1} \mathrm{d}\Phi \int_{\Theta_1}^{\frac{\pi}{2}} \frac{\sin^2\Theta}{R^2} R^2 \sin\Theta \mathrm{d}\Theta \\ &= 4G_i \nu_i^4 \arcsin \frac{x}{2R_x} \left[\frac{z}{2R_z} - \frac{1}{3} \left(\frac{z}{2R_z} \right)^3 \right] \,. \end{aligned} \tag{13.93}$$

where

$$\Phi_1 = \arcsin \frac{x}{2R_x}, \qquad \Theta_1 = \arccos \frac{z}{2R_z}$$

and R_x and R_z are the perpendicular distances from the dipole to the z and x sides, respectively. The z side of the cross-section rectangle is parallel with the dipole axis. The perpendicular distance between the dipole and the centre of the cross-section rectangle is given by $R_0 = \sqrt{R_z^2 - (z/2)^2}$.

For the near zone component $\langle W_{iR}\rangle$ of the electric field one may find

$$\begin{aligned}\langle W_{iR}\rangle = \int_D W_{iR}\mathrm{d}D &= 4g_i \int_0^{\Phi_1} \mathrm{d}\Phi \int_0^{\Theta_1} \frac{\cos^2\Theta}{R^2} R^2 \sin\Theta \mathrm{d}\Theta \\ &= \frac{4}{3} g_i \arcsin\frac{x}{2R_x}\left[1-\left(\frac{z}{2R_z}\right)^3\right] . \end{aligned} \tag{13.94}$$

The $\langle W_{i\Theta}\rangle$ component may be determined from Eq. (13.93) after substituting g_i for $4G_i\nu_i^4$.

In Eqs (13.93) and (13.94) we neglected the dependence of G_i and g_i on R, which is not constant in different points of the cross-section rectangle. As the dimensions x and z of the effective cross-section used in calculations are 5 nm and 10 nm, respectively, the error in the evaluation of the corresponding integrals for $R_0 > 10$ nm is smaller than about 10 %.

Eqs (13.93) and (13.94) are used to assess the capacity of the information channels between the oscillating units in the Fröhlich coherent vibration systems.

13.12 Interaction of Two Oscillators

Analysis of the interaction between oscillators can be useful for explanation of the important features of the interaction between biological macromolecules, structures, and between cells. Chapter 10 contains analysis of the interaction energy of two Fröhlich oscillators. We will present here an elementary treatment of the interaction between two identical oscillators with no damping and with the excitation at the initial time. Interacting oscillators are described by Eq. (10.2). Two identical oscillators in isotropic medium have the same masses, frequencies, and interaction coefficients, i.e. $m_1 = m_2 = m$, $\omega_1 = \omega_2 = \omega$, and $\gamma_1 = \gamma_2 = \gamma$. Therefore, they can be described by the equations

$$mq_1^{(2)} + m\omega^2 q_1 = \gamma q_2 , \quad mq_2^{(2)} + m\omega^2 q_2 = \gamma q_1 \tag{13.95}$$

where $q_1^{(2)}$ and $q_2^{(2)}$ are the second derivatives of the displacements q_1 and q_2 with respect to time, respectively.

From the first equation in (13.95) a relation for q_2 is found and substituted for q_2 in the second equation in (13.95). We obtain an equation whose solution is assumed in the exponential form $\exp(i\omega_\pm t)$. Substituting the exponential function we obtain the characteristic equation

$$(\omega_\pm)^4 - 2\omega^2(\omega_\pm)^2 + \omega^4 - \left(\frac{\gamma}{m}\right)^2 = 0$$

whose solution determines the frequencies of oscillations in the form

$$\omega_\pm^2 = \omega^2\left(1 \pm \frac{\gamma}{m\omega^2}\right) . \tag{13.96}$$

The general solution of the relations given in (13.95) can be written in the form

$$q_1 = A_1 \sin\omega_- t + B_1 \cos\omega_- t + A_1' \sin\omega_+ t + B_1' \cos\omega_+ t . \tag{13.97}$$

where A_1, B_1, A_1', B_1' are the amplitudes of oscillations.

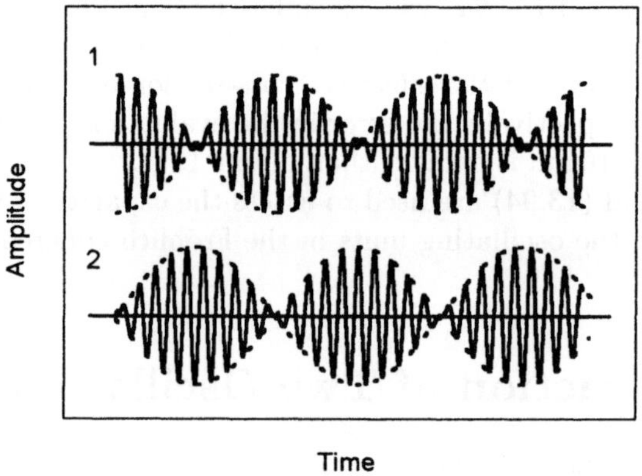

Fig 13.4 Oscillations in two identical interacting oscillators.

Superposition of oscillations with two slightly different frequencies leads to periodic variations of the amplitude. A series of periodically repeated beats appears in the form portrayed by Fig. 13.4. The frequency of these beats depends on the difference $\omega_+ - \omega_-$. The greater the difference the greater the frequency of beats.

Substituting Eq. (13.97) into the first equation in (13.95) and using Eq. (13.96) we get

$$q_2 = A_1 \sin\omega_- t + B_1 \cos\omega_- t - A_1' \sin\omega_+ t - B_1' \cos\omega_+ t . \tag{13.98}$$

Equations (13.97) and (13.98) show that two independent harmonic oscillations are present in each interacting oscillator. The oscillations with the frequency ω_- in both interacting oscillators are in phase and the oscillations with the frequency ω_+ have opposite phases. Therefore, the beats in the oscillators are shifted in phase.

We will assess the amplitudes of oscillations. Assuming that at the initial time we have

$$\frac{dq_1}{dt} = v\,, \quad q_1 = 0\,, \quad \frac{dq_2}{dt} = 0\,, \quad q_2 = 0 \tag{13.99}$$

we get from Eqs (13.97) and (13.98) that

$$A_1 = \frac{v}{2\omega_-}\,, \quad A_1' = \frac{v}{2\omega_+}\,, \quad B_1 = B_1' = 0\,. \tag{13.100}$$

As the energy of a harmonic oscillator ($q = A\sin\omega t$) is given by

$$W = \frac{1}{2}m\left(\frac{\partial q}{\partial t}\right)^2 + \frac{1}{2}m\omega^2 q^2 = \frac{1}{2}m\omega^2 A^2$$

we can conclude that the energy of oscillations at the frequency ω_- equals the energy at the frequency ω_+. Eq. (13.100) shows that $\omega_- A_1 = \omega_+ A_1'$. Fig. 13.4 shows oscillations in both interacting oscillators. Energy is transferred between them. When the energy in one oscillator has maximum value the energy in the other is zero and vice versa.

Using Eqs (13.97) and (13.100) the average energy of the interacting oscillator 1 is given by

$$W_1(R) = \frac{1}{2}m\omega_-^2 A_1^2 + \frac{1}{2}m\omega_+^2 (A_1')^2 = \frac{1}{4}mv^2 \tag{13.101}$$

where R stands for the finite distance between the interacting oscillators. A similar relation can be derived for $W_2(R)$ from Eqs (13.98) and (13.100). We find out that

$$W_2(R) = W_1(R)\,. \tag{13.102}$$

We compare the energy of an oscillator before interaction (at the distance $R \to \infty$) with its energy after interaction started. Solution of Eqs (13.95) for $\gamma = 0$ is given by

$$q_1 = A_1 \sin\omega t + B_1 \cos\omega t\,, \quad q_2 = A_2 \sin\omega t + B_2 \cos\omega t\,. \tag{13.103}$$

The initial conditions given by Eq. (13.99) are imposed on the non interacting oscillators too. From Eqs (13.103) we get

$$A_1 = \frac{v}{\omega}\,, \quad B_1 = A_2 = B_2 = 0\,. \tag{13.104}$$

Energy of the excited oscillator (denoted by the index 1) is given by

$$W_1(\infty) = \frac{1}{2}m\omega^2 A_1^2 = \frac{1}{2}mv^2 \tag{13.105}$$

and the non excited oscillator (denoted by the index 2) remains at rest, i.e.

$$W_2(\infty) = 0 \,. \tag{13.106}$$

The average energy of the excited oscillator given by Eq. (13.105) is twice greater than the energy of the same interacting oscillator (Eq. (13.101)). The total average energy of two interacting oscillators is equal to the energy of the oscillators before interaction.

The energy $W_1(R)$ in Eq. (13.101) (as well as $W_2(R)$, $W_1(\infty)$, $W_2(\infty)$ in Eqs (13.102), (13.105), (13.106), respectively) represents the free energy. The interaction energy I' is defined as a difference of the free energy $W(R) = W_1(R) + W_2(R)$ of both oscillators at a distance R (when the oscillators interact) and their free energy $W(\infty)$ (if exists) for $R \to \infty$. Therefore,

$$I' = W(R) - W(\infty) \tag{13.107}$$

and using Eqs (13.101), (13.102), and (13.105), (13.106) one gets from Eq. (13.107) that $I' = 0$.

If we assume that the excitation of the interacting oscillator 1 at the initial time is

$$\frac{dq_1}{dt} = 2v$$

then

$$A_1 = \frac{v}{\omega_-} \,, \quad A_1' = \frac{v}{\omega_+}$$

and its energy

$$\overline{W}_1(R) = mv^2$$

is twice greater than the total average energy of the interacting oscillators with the initial conditions given by (13.99), i.e. $\overline{W}_1(R) = 2\,[W_1(R) + W_2(R)]$. The interaction energy can be written in the form

$$I' = \overline{W}_1(R) - [W_1(\infty) + W_2(\infty)] \,.$$

where $W_1(\infty)$ and $W_2(\infty)$ hold for the excited oscillators.

Similar relations are used to assess the interaction energy of the Fröhlich systems.

13.13 Screening in Conductive Medium

A biological medium represents a conductive liquid. Within a region of nonvanishing conductivity no permanent distribution of free charges can be maintained by themselves. The applied electric field sets charges in motion and constitutes a current. The final steady state distribution of charges is such that the diffusion and the electric field are mutually balanced in dynamic equilibrium. The field around charged surfaces is screened by mobile charges which form a layer around them. The thickness of the screening layer and the time necessary for its building up are important parameters for assessment of the role of electric fields in biological medium. The conductivity of the liquid biological medium (of the cytosol and of the intercellular liquid) depends on free ions and charged particles, i.e. the conductivity is electrolytic.

The electrochemical potential μ describing the effects of both diffusion and electric fields may be expressed in the form [162]

$$\mu = \boldsymbol{R}T \ln a + z\boldsymbol{F}\varphi + \mu_0 \tag{13.108}$$

where $\boldsymbol{R}$ is the gas constant, $a = fc$ is the activity of ions (c is their concentration and f is the activity coefficient which equals 1 in very dilute solutions and 0.76 in the ordinary saline solution), z is the number of elementary charges per ion, $\boldsymbol{F}$ is the Faraday number, φ is the electric potential, z is the valency of an ion, and the additional term μ_0 depends on the nature of the solvent.

TABLE 4 Debey thickness δ

c_∞ (kmole m^{-3})	δ (nm)
10^{-4}	30.4
10^{-3}	9.6
10^{-2}	3.04
10^{-1}	0.96

TABLE 5 Relaxation time τ

σ (S)	τ (s)	ν (Hz)
10^{-4}	7.08×10^{-6}	1.4×10^{5}
10^{-3}	7.08×10^{-7}	1.4×10^{6}
10^{-2}	7.08×10^{-8}	1.4×10^{7}
10^{-1}	7.08×10^{-9}	1.4×10^{8}

In equilibrium the electrochemical potential in the whole space has the same value, i.e. $\mu(x) = \mu(x_0)$ at any point x. If μ_0 is constant then from Eq. (13.108) one gets

$$\mathrm{d}\mu = \boldsymbol{R}T\frac{\mathrm{d}a}{a} + z\boldsymbol{F}\mathrm{d}\varphi = 0 \tag{13.109}$$

which yields

$$a = a_0 \exp\left[-\frac{z\boldsymbol{F}}{\boldsymbol{R}T}(\varphi - \varphi')\right] . \tag{13.110}$$

If $\varphi' = 0$ and $c = c_\infty$ for $x \to \infty$ and f has the same value everywhere then Eq. (13.110) may be written as

$$c = c_\infty \exp\left(-\frac{z\boldsymbol{F}}{\boldsymbol{R}T}\varphi\right) . \tag{13.111}$$

If cations and anions are present in the solution the total charge density ρ is equal to the difference between the particular densities multiplied by the Faraday number $\boldsymbol{F}$. We get

$$\rho = \boldsymbol{F}(z_+ c_+ - z_- c_-) \tag{13.112}$$

where c_+ and c_- and are the concentrations of cations and anions, respectively. We will assume that valencies of cations and of anions equal one another, i.e. $z_+ = z_- = z$. For electrical neutrality $c_+(\infty) = c_-(\infty)$. From one-dimensional Poisson's equation

$$\frac{\mathrm{d}^2\varphi}{\mathrm{d}x^2} = -\frac{\rho}{\varepsilon'\varepsilon_0} \tag{13.113}$$

after substitution from Eqs (13.111) and (13.112) we get

$$\frac{\mathrm{d}^2\varphi}{\mathrm{d}x^2} = -\frac{z\boldsymbol{F}c_\infty}{\varepsilon'\varepsilon_0}\left[\exp\left(-\frac{z\boldsymbol{F}\varphi}{\boldsymbol{R}T}\right) - \exp\left(\frac{z\boldsymbol{F}\varphi}{\boldsymbol{R}T}\right)\right] . \tag{13.114}$$

For small values of the potential φ we get from Eq. (13.114)

$$\frac{\mathrm{d}^2\varphi}{\mathrm{d}x^2} = \frac{2\left(z\boldsymbol{F}\right)^2 c_\infty}{\boldsymbol{R}T\varepsilon'\varepsilon_0}\varphi = \kappa^2\varphi . \tag{13.115}$$

Solution of the Eq. (13.115) can be written in the form

$$\frac{\varphi}{\varphi_0} = \exp\left(-\kappa x\right) \tag{13.116}$$

and the reciprocal value of κ is reffered to as the Debye thickness defined by the relation

$$\delta = \frac{1}{\kappa} = \sqrt{\frac{\boldsymbol{R}T\varepsilon'\varepsilon_0}{2\left(z\boldsymbol{F}\right)^2 c_\infty}} . \tag{13.117}$$

The Debye thickness might be regarded as the "relaxation length". It determines the distance of effective screening. We will evaluate the Debye thickness for $T = 293\,\mathrm{K}$, $\varepsilon' = 80$, $z = 1$, and the standard values of $\boldsymbol{R}$, $\boldsymbol{F}$, and ε_0 ($\boldsymbol{R} = 8.31 \times 10^3\,\mathrm{JK^{-1}kmole^{-1}}$, $\boldsymbol{F} = 9.65 \times 10^7\,\mathrm{C\,kmole^{-1}}$, $\varepsilon_0 = 8.85 \times 10^{-12}\,\mathrm{Fm^{-1}}$). The values of δ for various concentrations of ions are given in Table 4. For a physiological concentration of ions the Debye thickness is of about several nanometers.

The charge screening layer is built up within a certain time interval. The time interval depends on the concentration and on the mobility of charge carriers (usually ions). The greater the concentration and the mobility the shorter the time interval. The equation of continuity can be used to assess the time interval which is necessary to achieve the steady state distribution. The equation of continuity

$$\frac{\partial \rho}{\partial t} = \frac{\sigma}{\varepsilon' \varepsilon_0} \rho \tag{13.118}$$

(where σ is the conductivity which depends on the mobility and on the concentration of the charge carriers) has solution

$$\rho = \rho_0 \exp \left(-\frac{\sigma}{\varepsilon' \varepsilon_0} t \right) = \rho_0 \exp \left(-\frac{t}{\tau} \right) . \tag{13.119}$$

The relaxation time τ is given be the relation

$$\tau = \frac{\varepsilon' \varepsilon_0}{\sigma} . \tag{13.120}$$

Table 5 shows the relaxation time τ and its reciprocal value $\nu = 1/\tau$ (the limit frequency) as a function of conductivity. The oscillating electric field can be screened if its frequency is lower than the limit value. In biological systems the limit frequency can be of about 10^8 MHz.

13.14 The Equilibrium Constant and Energy of Binding

The free energy for hydrolysis of the guanosine triphosphate (GTP) analogue guanylyl–(a,b)–methylene-diphosphonate (GMPCPP) to the GMPCP in the tubulin subunits in the microtubules was determined from the measurement of the equilibrium constant of the reaction [64,73]. The equilibrium constant (the affinity constant) employed as a measure of the binding strength between two entities (molecules) is determined from the rate of chemical reactions [1].

Equilibrium between entities A and B and their complex AB is maintained by a balance of two reactions – association and dissociation. The association reaction may be symbolically written as A + B $\rightarrow$ AB and similarly the dissociation reaction as AB $\rightarrow$ A + B. The association rate v_A is proportional to the product of the concentrations of the component A and of the component B. The dissociation rate v_D is proportional to the concentration of the complex AB. Therefore, v_A and v_D are given by the relations

$$v_A = k_{\rm on}[{\rm A}][{\rm B}], \qquad v_D = k_{\rm off}[{\rm AB}] \tag{13.121}$$

where $k_{\rm on}$ and $k_{\rm off}$ are the association and dissociation rate constants, respectively. The square brackets indicate the concentrations of the individual components A, B and of the complex AB. The larger the association rate constant and the smaller the dissociation rate constant the stronger the binding.

In equilibrium $v_A = v_D$ and, therefore,

$$k_{\text{on}}[\text{A}][\text{B}] = k_{\text{off}}[\text{AB}] \tag{13.122}$$

and the equilibrium constant K is given by the relation

$$K = \frac{k_{\text{on}}}{k_{\text{off}}} = \frac{[\text{AB}]}{[\text{A}][\text{B}]} . \tag{13.123}$$

The equilibrium constant given by Eq. (13.123) is related to the standard free energy change ΔG^{o} for the binding (the Gibbs free energy). The change of the free energy is given by the free energy of the complex AB from which the free energy of the components A + B is subtracted. We get

$$\Delta G^{\text{o}} = G(AB) - G(A + B) . \tag{13.124}$$

The equilibrium constant as a function of ΔG^{o} is given by

$$K = \frac{[\text{AB}]}{[\text{A}][\text{B}]} = \alpha \exp\left(-\Delta G^{\text{o}}/\boldsymbol{R}T\right) \quad \text{or} \quad \Delta G^{\text{o}} = -\boldsymbol{R}T \ln K/\alpha \tag{13.125}$$

where $\boldsymbol{R}$ is the gas constant and α is a coefficient adjusting the correct units of K (the numerical value of α equals 1 under standard conditions).

Table 6 shows the mutual dependence of K and ΔG^{o} for $T = 300\,\text{K}$ [1].

TABLE 6 The relationship between K and ΔG^{o}

K	ΔG^{o}	
(liters/mole)	(kcal/mole)	(kJ/mole)
10^4	−5.5	−23.0
10^2	−2.7	−11.5
1	0	0
10^{-2}	2.7	11.5
10^{-4}	5.5	23.0

The description of the association and dissociation rates is valid under standard conditions. The reactants must be brought together and collide to start the reaction. The product of the reaction must receive certain energy to dissociate. We may assume that under standard conditions the thermal energy and the random thermal motion are responsible for initiation of these reactions.

Bibliography

[1] ALBERTS B., BRAY D., LEWIS J., RAFF M., ROBERTS K., and WATSON J. D. *Molecular Biology of the Cell (third edition)*. Garland Publishing, New York, London, 1994.

[2] CLAESSON P.M. The Surface Force Technique: Interaction Between Surfactant Bilayers and Between Protein Layers. In BONGRAND P., CLAESSON P.M., and CURTIS A.S.G., editors, ***Studying Cell Adhesion***, pages 15–35. Springer-Verlag, Berlin, Heidelberg, New York, 1994.

[3] PARSEGIAN V.A. Harness the Hubris: Useful Things Physicists Could do in Biology. *Physics Today*, 50:23–27, 1997.

[4] NICOLIS G. and PRIGOGINE I. ***Self-Organization in Nonequilibrium Systems***. J. Wiley & Sons, New York, London, Sydney, Toronto, 1977.

[5] HOPPE W., LOHMANN W., MARKL H., and ZIEGLER H. ***Biophysik***. Springer-Verlag, Berlin, Heidelberg, New York, 1982.

[6] FRÖHLICH H. Quantum mechanical concepts in biology. In MAROIS M., editor, ***Theoretical Physics and Biology***, pages 13–22. North-Holland Pub. Comp., Amsterdam, 1969.

[7] FRÖHLICH H. Bose Condensation of Strongly Excited Longitudinal Electric Modes. *Phys. Lett.*, 26A:402–403, 1968.

[8] FRÖHLICH H. Long-Range Coherence and Energy Storage in Biological Systems. *Int. J. Quant. Chem.*, II:641–649, 1968.

[9] POPP F.A. Electromagnetism and Living System. In HO M.-W., POPP F.A., and WARNKE U., editors, ***Bioelectrodynamics and Biocommunication***. World Scientific, Singapore, New Jersey, London, Hong Kong, 1994.

[10] VOS M. H., RAPPAPORT F., LAMBRY J.-CH., BRETON J., and MARTIN J. L. Visualization of coherent nuclear motion in a membrane protein by femtosecond spectroscopy. *Nature*, 363:320–325, 1993.

[11] BLUMENFELD L.A. ***Physics of Bioenergetic Processes***. Springer-Verlag, Berlin, Heidelberg, New York, 1983.

[12] ISRAELACHVILI J., LECKBAND D., SCHMITT F.-J., ZASADZINSKI J., WALKER S., and CHIRUVOLU S. Direct Measurements of Specific Ligand-Receptor Interactions Between Model Membrane Surfaces. In BONGRAND P., CLAESSON P.M., and CURTIS A.S.G., editors, ***Studying Cell Adhesion***, pages 37–49. Springer-Verlag, Berlin, Heidelberg, New York, 1994.

[13] ROWLANDS S., SEWCHAND L.S., LOVLIN R.E., BECK J.S., and ENNS E.G. A Fröhlich Interaction of Human Erythrocytes. *Phys. Lett.*, 82A:436–438, 1981.

[14] ROWLANDS S., SEWCHAND L.S., and ENNS E.G. Further Evidence for a Fröhlich Interaction of Erythrocytes. *Phys. Lett.*, 87A:256–260, 1982.

[15] ROWLANDS S. and SEWCHAND L.S. A quantum mechanical interaction of human erythrocytes. *Canad. J. Physiol. Pharmacol.*, 60:52–59, 1982.

[16] SEWCHAND L.S., ROBERTS D., and ROWLANDS S. Transmission of the Quantum Interaction of Erythrocytes. *Cell. Biophys.*, 4:253–259, 1982.

[17] ROWLANDS S. Coherent Excitation in Blood. In FRÖHLICH H. and KREMER F., editors, *Coherent Excitations in Biological Systems*, pages 145–161. Springer-Verlag, Berlin, Heidelberg, New York, 1983.

[18] ROWLANDS S., SEWCHAND L.S., and SKIBO L. Conversion of Albumin into a Transmitter of the Ultra Long-Range Interaction of Human Erythrocytes. *Cell Biophys.*, 5:197–203, 1983.

[19] ROWLANDS S., SKIBO L., EISENBERG C. P., and SEWCHAND L.S. Triple Contractils. *J. Biol. Phys.*, 12:31–32, 1984.

[20] SEWCHAND L.S., MASRI M.A., FRITZ O.G., BOYD N.G., and ROWLANDS S. Fibrinogen and the Ultralong-Range Interaction of Human Erythrocytes. *Cell Biophys.*, 6:215–221, 1984.

[21] ROWLANDS S. Quantum Mechanical Coherence in Human Red Blood Cells. *J. Statist. Phys.*, 39:543–549, 1985.

[22] ROWLANDS S. The Interaction of Living Red Blood Cells. In FRÖHLICH H., editor, *Biological Coherence and Response to External Stimuli*, pages 171–191. Springer-Verlag, Berlin, Heidelberg, New York, 1988.

[23] ROWLANDS S. Rouleau formation of erythrocytes as evidence of Fröhlich's mechanism. *Neural Network World*, 4:339–356, 1994.

[24] POKORNÝ J., JANDOVÁ A., ŠORFOVÁ J., KOBILKOVÁ J., TROJAN S., COSTATO M., and MILANI M. Leukocytes of cancer patients give evidence of fundamental physical forces: a pathway for a new view. *Laser and Technology*, 6(1–2):15–23, 1996.

[25] JANDOVÁ A., KOBILKOVÁ J., NEDBALOVÁ M., ŠORFOVÁ J., DOHNALOVÁ A., TROJAN S., and POKORNÝ J. Study of the interaction forces in human CD_4Ly. *Bioelectrochem. and Bioenergetics*, 41:181–185, 1996.

[26] CORBÍ A.L. *Leukocyte Integrins: Structure, Expression and Function.* Springer-Verlag, New York, Berlin, Heidelberg, 1996.

[27] POHL H.A. Oscillating Fields about Growing Cells. *Int. J. Quant. Chem.: Quant. Biol. Symp.*, 7:411–431, 1980.

[28] POHL H.A. Do Cells in a Reproductive State Exhibit a Fermi-Pasta-Ulam-Fröhlich Resonance and Emit Electromagnetic Radiation? *J. Biol. Phys.*, 8:45–75, 1980.

[29] POHL H.A., BRADEN T., ROBINSON S., PICLARDI J., and POHL D.G. Life Cycle Alterations of the Micro-Dielectrophoretic Effects of Cells. *J. Biol. Phys.*, 9:133–154, 1981.

[30] ROY S.C., BRADEN T., and POHL H.A. Possibility of Existence of Pseudoferroelectric State in Cells: Some Experimental Evidence. *Phys. Lett.*, 83A:142–143, 1981.

[31] LAFON E. and POHL H.A. Critical Fields in Cell Fusion and Dielectrophoresis. *J. Biol. Phys.*, 10:43–50, 1982.

[32] POHL H.A. Natural Oscillating Fields of Cells. In FRÖHLICH H. and KREMER F., editors, *Coherent Excitations in Biological Systems*, pages 199–210. Springer-Verlag, Berlin, Heidelberg, New York, 1983.

[33] POHL H.A. *Dielectrophoresis.* Cambridge Univ. Press, London, New York, Melbourne, 1978.

[34] SAUER F.A. Forces on Suspended Particles in the Electromagnetic Field. In FRÖHLICH H. and KREMER F., editors, *Coherent Excitations in Biological Systems*, pages 134–144. Springer-Verlag, Berlin, Heidelberg, New York, 1983.

[35] JOOS G. *Theoretical Physics*. G. E. Stechert & Co., New York, 1934.

[36] PRIGOGINE I. Structure, Dissipation and Life. In MAROIS M., editor, *Theoretical Phys. and Biol.*, pages 23–52. North-Holland, 1969.

[37] NICOLIS J.S. *Dynamics of Hierarchical Systems. An Evolutionary Approach*. Springer-Verlag, Berlin, Heidelberg, New York, 1986.

[38] LAM L. *Introduction to Nonlinear Physics*. Springer-Verlag, New York, Berlin, Heidelberg, 1997.

[39] HAKEN H. Cooperative phenomena in systems far from thermal equilibrium and nonphysical systems. *Rev. Mod. Phys.*, 47(1):67–121, 1975.

[40] HAKEN H. *Synergetics*. Springer-Verlag, Berlin, Heidelberg, New York, 1977.

[41] HAKEN H. *Evolution of Order and Chaos*. Springer-Verlag, Berlin, Heidelberg, New York, 1982.

[42] HAKEN H. *Information and Self-Organization*. Springer-Verlag, Berlin, Heidelberg, New York, 1988.

[43] MIKHAILOV A.S. *Foundation of Synergetics I*. Springer-Verlag, Berlin, Heidelberg, New York, 1994.

[44] ALLIGOOD K.T., SAUER T.D., and YORKE J.A. *Chaos*. Springer-Verlag, New York, Berlin, Heidelberg, 1997.

[45] KITTEL C. *Introduction to Solid State Physics*. J. Wiley & Sons, Chapman & Hall, New York, London, 1956.

[46] DEKKER A.J. *Solid State Physics*. Prentice-Hall, Englewood Cliffs, 1957.

[47] ANSELM A.I. *Vvedenie v teoriju poluprovodnikov (Introduction to the Theory of Semiconductors)*. Gosud. Izdat. Mat.-Fyz. Liter., Moscow, Leningrad, 1962.

[48] ZIMAN J.M. *Models of Disorder*. Cambridge Univ. Press, Cambridge, London, New York, Melbourne, 1979.

[49] AMOS L. A. and KLUG A. Arrangement of Subunits in Flagellar Microtubules. *J. Cell. Sci.*, 14:523–549, 1974.

[50] TUSZYŃSKI J.A., TRPISOVA B., and SEPT D. From erratic to coherent behaviour in the assembly of microtubules. *Neural Network World*, 5:675–688, 1995.

[51] MANDELKOW E., MANDELKOW E.-M., HOTANI H., HESS B., and MÜLLER S.C. Spatial Patterns from Oscillating Microtubules. *Science*, 246:1291–1293, 1989.

[52] MELKI R., CARLIER M.-F., and PANTALONI D. Direct Evidence for GTP and GDP–P_i Intermediates in Microtubule Assembly. *Biochemistry*, 29:8921–8932, 1990.

[53] MANDELKOW E.-M., MANDELKOW E., and MILLIGEN R.A. Microtubule Dynamics and Microtubule Caps: A Time-resolved Cryo-Electron Microscopy Study. *J. Cell Biol.*, 114:977–991, 1991.

[54] RODIONOV V.I., LIM S.-S., GELFAND V.I., and BORISY G.G. Microtubule Dynamics in Fish Melanophores. *J. Cell Biol.*, 126:1455–1464, 1994.

[55] MELKI R., CARLIER M.-F., PANTALONI D., and TIMASHEFF S.N. Cold Depolymerization of Microtubules to Double Rings: Geometric Stabilization of Assemblies. *Biochem.*, 28:9143–9152, 1989.

[56] BAAS P.W., BLACK M.M., and BANKER G.A. Changes of Microtubule Polarity Orientation during The Development of Hippocampal Neurons in Culture. *J. Cell Biol.*, 109:3085–3094, 1989.

[57] SATARIĆ M.V., TUSZYŃSKI J.A., HAMEROFF S., and ŽAKULA R.B. Microtubules and their role in neuromolecular computing. *Neural Network World*, 4:281–294, 1994.

[58] TUSZYŃSKI J.A., HAMEROFF S., SATARIĆ M.V., TRPISOVA B., and NIP M.L.A. Ferroelectric Behavior in Microtubule Dipole Lattices: Implications for Information Processing, Signaling and Assembly/Disassembly. *J. theor. Biol*, 174:371–380, 1995.

[59] MOGENSEN M.M., TUCKER J.B., and STEBBINGS H. Microtubule Polarities Indicate that Nucleation and Capture of Microtubules Occurs at Cell Surfaces in *Drosophila*. *J. Cell Biol.*, 108:1445–1452, 1989.

[60] PERÄNEN J., AUVINEN P., VIRTA H., WEPF R., and SIMONS K. Rab8 Promotes Polarized Membrane Transport through Reorganization of Actin and Microtubules in Fibroblasts. *J. Cell Biol.*, 135:153–167, 1996.

[61] BILLINGSLEY M.L. and KINCAID R.L. Regulated phoshorylation and dephosphorylation of tau protein: effects on microtubule interaction, intercellular trafficking and neurodegeneration. *Biochem. J.*, 323:577–591, 1997.

[62] CAPLOW M. and SHANKS J. Induction of Microtubule Catastrophe by Formation of Tubulin–GDP and Apotubulin Subunits at Microtubule Ends. *Biochemistry*, 34:15732–15741, 1995.

[63] HYMAN A.A. and MITCHISON T.J. Modulation of Microtubule Stability by Kinetochores In Vitro. *J. Cell Biol.*, 110:1607–1616, 1990.

[64] CAPLOW M., RUHLEN R.L., and SHANKS J. The Free Energy for Hydrolysis of a Microtubule-Bound Nucleotide Triphosphate Is Near Zero: All of the Free Energy Is Stored in the Microtubule Lattice. *J. Cell Biol.*, 127:779–788, 1994.

[65] SATARIĆ M.V., ŽAKULA R.B., ZEKOVIĆ S., POKORNÝ J., and FIALA J. The change of microtubule length caused by endogenous AC fields in cell. *BioSystems*, 39:127–133, 1996.

[66] SATARIĆ M.V., POKORNÝ J., FIALA J., ŽAKULA R.B., and ZEKOVIĆ S. Microtubules in interactions with endogenous d.c. and a.c. fields in living cells. *Bioelectrochem. and Bioenergetics*, 41(1):53–58, 1996.

[67] KÄS J., STREY H., TANG J.X., FINGER D., EZZELL R., SACKMANN E., and JANMEY P.A. F-actin, a Model Polymer for Semiflexible Chains in Dilute, Semidilute, and Liquid Crystalline Solutions. *Biophys. J.*, 70:609–625, 1996.

[68] SATO M., SCHWARTZ W.H., SELDEN S. CH., and POLLARD T.D. Mechanical Properties of Brain Tubulin and Microtubules. *J. Cell. Biol.*, 106:1205–1211, 1988.

[69] JANMEY P.A., EUTENEUER U., TRAUB P., and SCHLIWA M. Viscoelastic Properties of Vimentin Compared with Other Filamentous Biopolymer Networks. *J. Cell Biol.*, 113:155–160, 1991.

[70] LETERRIER J.F., KÄS J., HARTWIG J., VEGNERS R., and JANMEY P.A. Mechanical Effects of Neurofilament Cross-bridges. *The J. Biol. Chem.*, 271:15687–15694, 1996.

[71] JANMEY P.A. Coping with Cellular Stress: The Mechanical Resistance of Porous Protein Networks. *Biophys. J.*, 71:3–7, 1996.

[72] MACKINTOSH F.C., KÄS J., and JANMEY P.A. Elasticity of Semiflexible Biopolymer Networks. *Phys. Rev. Lett.*, 75:4425–4428, 1995.

[73] CAPLOW M. and SHANKS J. Evidence that a Single Monolayer Tubulin–GTP Cap Is Both Necessary and Sufficient to Stabilize Microtubules. *Molec. Biol. Cell*, 7:663–675, 1996.

[74] SATARIĆ M.V., TUSZYŃSKI J.A., and ŽAKULA R.B. Kinklike excitations as an energy-transfer mechanism in microtubules. *Phys. Rev. E*, 48:589–597, 1993.

[75] FRÖHLICH H. The Biological Effects of Microwaves and Related Questions. In *Advances in Electronics and Electron Phys.*, volume 53, pages 85–152. 1980.

[76] MAVROMATOS N.E. and NANOPOULOS D.V. On Quantum Mechanical Aspects of Microtubules. *Int. J. Mod. Phys.*, 1998.

[77] JIBU M., HAGAN S., HAMEROFF S. R., PRIBRAM K. H., and YASUE K. Quantum optical coherence in cytoskeletal microtubules: implications for brain function. *BioSystems*, 32:195–209, 1994.

[78] KITTEL C. *Quantum Theory of Solids*. J. Wiley & Sons, New York, London, 1963.

[79] FRÖHLICH H. Long-Range Coherence in Biological Systems. *Rivista del Nuovo Cim.*, 7:399–418, 1977.

[80] FRÖHLICH H. Kogerentnyje vozbuzhdenija v biologicheskikh systemakh (Coherent excitations in biological systems). *Biofizika*, XXII:743–744, 1977.

[81] FRÖHLICH H. Coherent Electric Vibrations in Biological Systems and Cancer Problem. *IEEE Trans.*, MTT-26:613–617, 1978.

[82] FRÖHLICH H., PELZER H., and ZIENAU S. Properties of Slow Electrons in Polar Materials. *Phil. Mag.*, 41:221–242, 1950.

[83] BHAUMIK D., BHAUMIK K., and DUTTA-ROY B. On the Possibility of "Bose Condensation" in the Excitation of Coherent Modes in Biological Systems. *Phys. Lett.*, 56A:145–148, 1976.

[84] BHAUMIK D., BHAUMIK K., and DUTTA-ROY B. A Microscopic Approach to the Fröhlich Model of Bose Condensation of Phonons in Biological Systems. *Phys. Lett.*, 59A:77–80, 1976.

[85] WU T.-M. and AUSTIN S. Bose Condensation in Biosystems. *Phys. Lett.*, 64A:151–152, 1977.

[86] WU T.-M. and AUSTIN S. Cooperative Behaviour in Biological Systems. *Phys. Lett.*, 65A:74–76, 1978.

[87] WU T.-M. and AUSTIN S. Bose-Einstein Condensation in Biological Systems. *J. theor. Biol.*, 71:209–214, 1978.

[88] WU T.-M. and AUSTIN S. Biological Bose Condensation and the Time Threshold for Biological Effects. *Phys. Lett.*, 73A:266–268, 1979.

[89] MILLS R.E. Asymptotic Solutions of the Fröhlich Steady-State Model. *Phys. Lett.*, 74A:278–280, 1979.

[90] MILLS R.E. Fröhlich Nonthermal Excitations in Biological Systems. *Phys. Lett.*, 91A:91–94, 1982.

[91] MILLS R.E. Comment on Yushina's Denial of the Existence of Fröhlich Nonthermal Excitations. *Phys. Lett.*, 93A:314–316, 1983.

[92] MILLS R.E. Fröhlich's model of nonthermal excitations in biological systems. *Phys. Rev.*, A28:379–394, 1983.

[93] MILLS R.E. Steady states in the extended Fröhlich model of far-from-equilibrium energy transport. *Phys. Rev.*, A41:2196–2199, 1990.

[94] MILLS R.E. Comment on "Relationship between Fröhlich and Davydov models of biological order". *Phys. Rev.*, A43:3176–3178, 1991.

[95] MILLS R.E. The Fröhlich rate equations. *Neural Network World*, 4:245–254, 1994.

[96] TUSZYŃSKI J.A. Generalization of Fröhlich's Thermodynamic Model of the Living State Phase Transition. *Phys. Lett.*, 108A:177–182, 1985.

[97] DEL GIUDICE E., DOGLIA S., MILANI M., and VITIELLO G. A Quantum Field Theoretical Approach to the Collective Behaviour of Biological Systems. *Nuclear Phys.*, B251:375–400, 1985.

[98] DEL GIUDICE E., DOGLIA S., and MILANI M. Self-Focusing of Fröhlich Waves and Cytoskeleton Dynamics. *Phys. Lett.*, 90A:104–106, 1982.

[99] BOLTERAUER H., TUSZYŃSKI J.A., and SATARIĆ M.V. Fröhlich and Davydov regimes in the dynamics of dipolar oscillations of biological membranes. *Phys Rev.*, A44:1366–1381, 1991.

[100] TUSZYŃSKI J.A. and PAUL R. Reply to "Comment on 'Relationship between Fröhlich and Davydov models of biological order'". *Phys. Rev.*, A43:3179, 1991.

[101] CHATTERJEE R., TUSZYŃSKI J.A., and PAUL R. Comments on the Bose Condensation of Phonons in the Biological Systems. *Int. J. Quant. Chem.*, XXIII:709–712, 1983.

[102] BOLTERAUER H. and LUDWIG L.A. Bose condensation for the Wu-Austin Hamiltonian without pumping. *Phys. Rev.*, E47:2122–2127, 1993.

[103] BOLTERAUER H. and LUDWIG L.A. Difficulties in the theoretical foundations of the Fröhlich's model. *Neural Network World*, 4:255–268, 1994.

[104] ABRIKOSOV A.A., GORKOV L.P., and DZYALOSHINSKII I.Y. *Quantum Field Theoretical Method in Statistical Physics.* Pergamon, Oxford, London, Edinburgh, New York, 1965.

[105] LIVSHITS M.A. K voprosu ob uchastiji kogerentnykh fononov v biologicheskikh procesakh (To the question of participation of coherent phonons in biological processes). *Biofizika*, XVII:694–695, 1972.

[106] LIVSHITS M.A. O vozmozhnosti Boze-kondensaciji v biologicheskikh sistemakh (On the possibility of Bose-condensation in biological systems). *Biofizika*, XXII:744–746, 1977.

[107] POKORNÝ J., VACEK K., and FIALA J. The Fröhlich Kinetic Equation with Additional Terms. *Czech. J. Phys.*, B36:1443–1450, 1986.

[108] POKORNÝ J., VACEK K., and FIALA J. Multiple Quantum Terms in the Fröhlich Kinetic Equation. *Phys. Lett.*, 114A:339–340, 1986.

[109] MANLEY J.M. and ROWE H.E. Some General Properties of Nonlinear Elements – Part I. General Energy Relations. *Proc. IRE*, 44:904–913, 1956.

[110] YUSHINA M.YA. Derivation of Kinetic Equation in Fröhlich Anharmonic Model. *Phys. Lett.*, 91A:372–374, 1982.

[111] FRÖHLICH H. Conditions for Coherent Excitations in Biological Systems. *Phys. Lett.*, 93A:105–106, 1982.

[112] POKORNÝ J. Computer Analysis of Coherent Electric Vibrations in Biological Systems. *Czech. J. Phys.*, B32:928–935, 1982.

[113] POKORNÝ J. Multiple Fröhlich Coherent States in Biological Systems: Computer Simulation. *J. theor. Biol.*, 98:21–27, 1982.

[114] POKORNÝ J., FIALA J., and VACEK K. Computer simulation of the Fröhlich coherent vibrations. *Czech. J. Phys.*, 41:1289–1294, 1991.

[115] WOLFRAM S. *Mathematica*. Addison-Wesley, Reading, Menlo Park, New York, 1991.

[116] WADA A. The α-Helix as an electric Macro-Dipole. *Adv. Biophys.*, 9:1–63, 1976.

[117] WADA A. and NAKAMURA H. Nature of the charge distribution in proteins. *Nature*, 293:757–758, 1981.

[118] WADA A., NAKAMURA H., and SAKAMOTO T. Nature of the Charge Distribution in Proteins. II. Effect of Atomic Partial Charges on Ionic Charges. *J. Phys. Soc. Jap.*, 54:4042–4046, 1985.

[119] NAKAMURA H. and WADA A. Nature of the Charge Distribution in Proteins. III. Electric Multipole Structures. *J. Phys. Soc. Jap.*, 54:4047–4052, 1985.

[120] WALTON G.A. *Polypeptides and protein structure*. Elsevier, New York, 1981.

[121] DAUXOIS T. and PEYRARD M. Dynamics of Breather Modes in a Nonlinear 'Helicoidal' Model of DNA. In REMOISSENET M. and PEYRARD M., editors, *Nonlinear Coherent Structures in Phys. and Biol., Proc. VIIth Interdiscipl. Workshop, Dijon, France, 4–6 June*, pages 79–86. Springer-Verlag, Berlin, Heidelberg, New York, 1991.

[122] GRUNDLER W., KEILMANN F., and FRÖHLICH H. Resonant Growth Rate Response of Yeast Cells Irradiated by Weak Microwaves. *Phys. Lett.*, 62A:463–466, 1977.

[123] GRUNDLER W. and KEILMANN F. Nonthermal Effects of Millimeter Microwaves on Yeast Growth. *Zeitschrift für Naturforschung*, 33c:15–22, 1978.

[124] GRUNDLER W., KEILMANN F., PUTTERLIK V., SANTO L., STRUBE D., and ZIMMERMANN I. Nonthermal Resonant Effects of 42-GHz Microwaves on the Growth of Yeast Cultures. In FRÖHLICH H. and KREMER F., editors, *Coherent Excitations in Biological Systems*, pages 21–37. Springer-Verlag, Berlin, Heidelberg, New York, 1983.

[125] GRUNDLER W. and KEILMANN F. Biological Effects of RF and MW Energy at Molecular and Cellular Level. In GRANDOLFO M., MICHAELSON S.M., and RINDI A., editors, *Biol. Effects and Dosimetry of Nonionizing Radiation*, pages 299–318. Plenum Pub. Corp., 1983.

[126] GRUNDLER W. and KEILMANN F. Sharp Resonance in Yeast Growth Prove Nonthermal Sensitivity of Microwaves. *Phys. Rev. Lett.*, 51:1214–1216, 1983.

[127] KEILMANN F. Experimental RF and MW Resonant Nonthermal Effects. In GRANDOLFO M., MICHAELSON S.M., and RINDI A., editors, *Biol. Effects and Dosimetry of Nonionizing Radiation*, pages 283–297. Plenum Pub. Corp., 1983.

[128] GOS P., EICHER B., KOHLI J., and HEYER W.-D. Extremly High Frequency Electromagnetic Fields at Low Power Density Do Not Affect the Division of Exponential Phase Saccharomyces cerevisiae Cells. *Bioelectromagnetics*, 18:142–155, 1997.

[129] POKORNÝ J. and FIALA J. Condensed Energy and Interaction Energy of Fröhlich Systems. *Neural Network World*, 4:299–313, 1994.

[130] POKORNÝ J. and FIALA J. Heat Bath Coupling Effects in Coherent Vibration Systems. *Europhys. Lett.*, 19:729–734, 1992.

[131] VOJTÁŠEK S. and JANÁČ K. *Solution of Non-Linear Systems*. Academia, Prague, Czechoslovakia, 1969.

[132] HAKEN H. General Discussion. In MAROIS M., editor, *Theoretical Phys. and Biol.*, pages 94–97. North-Holland, 1969.

[133] SEWELL G.L. Quantum model of biological coherence and chaos. *Neural Network World*, 5:725–732, 1995.

[134] MANDEL L. and Wolf E. *Optical coherence and quantum optics*. Cambridge Univ. Press, Cambridge, 1995.

[135] SALEH B.E.A. and TEICH M.C. *Fundamentals of Photonics*. J. Wiley & Sons, New York, 1991.

[136] PAPOULIS A. *Probabily, Random Variables, and Stochastic Processes*. McGraw-Hill, New York, 1965.

[137] BECK A.H.W. *Statistical Mechanics, Fluctutions, and Noise*. E. Arnold, London, 1976.

[138] FRÖHLICH H. Evidence for Bose Condensation-Like Excitation of Coherent Modes in Biological Systems. *Phys. Lett.*, 51A:21–22, 1975.

[139] ROWE H.E. Some General Properties of Nonlinear Elements. II. Small Signal Theory. *Proc. IRE*, 46:850–860, 1958.

[140] BLACKWELL L.A. and KOTZEBUE K.L. *Semiconductor-Diode Parametric Amplifiers*. Prentice-Hall, Englewood Cliffs, 1961.

[141] PENFIELD P. and RAFUSE R.P. *Varactor Applications*. M.I.T. Press, Cambridge, Massachusetts, 1962.

[142] HYDE F.J. and TUCKER D.G. Parametric Amplifiers: Static and Dynamic Inductance and Capacitance and their Significance in the Non-Linear Time-Varying Approaches. *J. Brit. I.R.E.*, 25(4):353–355, 1963.

[143] TUCKER D.G. Circuits with Time-Varying Parameters. *J. Brit. I.R.E.*, 25(3):263–271, 1963.

[144] KUDREWICZ J. Power-Frequency Relations in Nonlinear Two-Poles. *Proc. IEEE*, 51:1599–1605, 1963.

[145] POKORNÝ J., JELÍNEK F., and TRKAL V. Information Transfer in Biological Systems with Endogenous Electromagnetic Field. In *EBEA'96 - 3rd Int. Congress Europ. Bio Electromag. Ass., Feb. 29–March 3, Nancy, France, Abstract Book*, 1996.

[146] STRATTON J.A. *Electromagnetic Theory*. McGraw-Hill, New York, London, 1941.

[147] WILKENING G.M, BARNES F.S., BOWMAN R.R., GUY A.W., ILLINGER K.H., JOHNSON C.C., ROSENTHAL S.W., and SUSSKIND CH. Radiofrequency Electromagnetic Fields. Technical Report 67, Natl. Council on Radiation Protection and Measurements, Washigton, D.C., USA, 1981.

[148] POKORNÝ J. Coherent Excitations and Electromagnetic Field Generated by Living Cells. In FIALA J. and POKORNÝ J., editors, *Proc Int. Seminar Biophysical Aspects of Cancer, July 6-9, Prague, Czechoslovakia*, pages 102–110. Charles University, 1987.

[149] SEWCHAND L.S. and ROWLANDS S. Specificity of the Fröhlich Interaction of Erythrocytes. *Phys. Lett.*, 93A:363–364, 1983.

[150] LADIK J. and FÖRNER W. *The Beginnings of Cancer in the Cell*. Springer-Verlag, Berlin, Heidelberg, New York, 1994.

[151] HARTLEY R.V.L. Transmission of information. *Bell Syst. Tech. J.*, 7:535–563, 1928.

[152] SHANNON C.E. A Mathematical Theory of Communication. *Bell Syst. Tech. J.*, 27:379–423, 1948.

[153] SHANNON C.E. A Mathematical Theory of Communication. *Bell Syst. Tech. J.*, 27:623–656, 1948.

[154] OLIVER B.M., PIERCE J.R., and SHANNON C.E. The philosophy of P.C.M. *Proc. IRE*, 36:1324–1331, 1948.

[155] SHANNON C.E. Communication in the presence of noise. *Proc. IRE*, 37:10–21, 1949.

[156] TULLER W. G. Theoretical limitations on the rate of transmission of information. *PIRE*, 37:468–478, 1949.

[157] POKORNÝ J. and FIALA J. Heat Bath Coupling Effects on Interaction Between Fröhlich Vibration Systems. *Czech J. Phys.*, 43:67–78, 1994.

[158] YU F.T.S. *Optics and Information Theory*. J. Wiley & Sons, New York, 1976.

[159] KOLMOGOROV A.N. *Teorija informacii i teorija algoritmov (Theory of information and theory of algorithms)*. Nauka, Moscow, 1987.

[160] FRÖHLICH H. Selective Long Range Dispersion Forces between Large Systems. *Phys. Lett.*, 39A:153–154, 1972.

[161] POKORNÝ J. Fröhlich coherent states: Implications for interaction in biological systems and information transfer. *Neural Network World*, 5:831–847, 1995.

[162] KOTYK A. and JANÁČEK K. *Membrane Transport*. Academia, Prague, 1977.

[163] NURSE P. Universal control mechanism regulating onset of M-phase. *Nature*, 344:503–508, 1990.

[164] NASMYTH K. Control of the yeast cell cycle by the Cdc28 protein kinase. *Current Opinion in Cell Biol.*, 5:166–179, 1993.

[165] JOHNSON L.N. and BARFORD D. The Effects of Phosphorylation on the Structure and Function of Proteins. In ENGELMAN D.M., CANTOR CH.R., and POLLARD T.D., editors, *Annu. Rev. Biophys. Biomol. Struct.*, volume 22, pages 199–232. Annual Reviews, 1993.

[166] HUNTER T. The Proteins of Oncogenes. *Sci. American*, 251(2):60–69, 1984.

[167] HUNTER T. and PINES J. Cyclins and Cancer. *Cell*, 66:1071–1074, 1991.

[168] LEVINSON A.D., OPPERMANN H., LEVINTOW L., VARMUS H.E., and BISHOP J.M. Evidence that the Transforming Gene of Avian Sarcoma Virus Encodes a Protein Kinase Associated with a Phosphoprotein. *Cell*, 15:561–572, 1978.

[169] HUNTER T. and SEFTON B.M. Transforming gene product of Rous sarcoma virus phosphorylates tyrosine. *Proc. Nat. Acad. Sci. USA (Biochemistry)*, 77:1311–1315, 1980.

[170] HUNTER T. and COOPER J.A. Epidermal Growth Factor Induces Rapid Tyrosine Phosphorylation of Proteins in A431 Human Tumor Cells. *Cell*, 24:741–752, 1981.

[171] ABERCROMBIE M. and AMBROSE E.J. The Surface Properties of Cancer Cells. A Review. *Cancer Res.*, 22:525–548, 1962.

[172] NETLAND P.A. and ZETTER B.R. Organ-Specific Adhesion of Metastic Tumor Cells in Vitro. *Science*, 224(4653):1113–1114, 1984.

[173] BRODT P. *Cell Adhesion and Invasion in Cancer Metastasis*. Springer-Verlag, Heidelberg, 1996.

[174] RAIMES S. *Many-Electron Theory*. North-Holland, Amsterdam, London, 1972.

Index

MIX
Papier aus verantwortungsvollen Quellen
Paper from responsible sources
FSC® C105338

FSC
www.fsc.org

If you have any concerns about our products,
you can contact us on
ProductSafety@springernature.com

In case Publisher is established outside the EU,
the EU authorized representative is:
Springer Nature Customer Service Center GmbH
Europaplatz 3, 69115 Heidelberg, Germany

Printed by Libri Plureos GmbH
in Hamburg, Germany